Research for Development

The series Research for Development serves as a vehicle for the presentation and dissemination of complex research and multidisciplinary projects. The published work is dedicated to fostering a high degree of innovation and to the sophisticated demonstration of new techniques or methods.

The aim of the Research for Development series is to promote well-balanced sustainable growth. This might take the form of measurable social and economic outcomes, in addition to environmental benefits, or improved efficiency in the use of resources; it might also involve an original mix of intervention schemes.

Research for Development focuses on the following topics and disciplines:

Urban regeneration and infrastructure, Info-mobility, transport, and logistics, Environment and the land, Cultural heritage and landscape, Energy, Innovation in processes and technologies, Applications of chemistry, materials, and nanotechnologies, Material science and biotechnology solutions, Physics results and related applications and aerospace, Ongoing training and continuing education.

Fondazione Politecnico di Milano collaborates as a special co-partner in this series by suggesting themes and evaluating proposals for new volumes. Research for Development addresses researchers, advanced graduate students, and policy and decision-makers around the world in government, industry, and civil society.

THE SERIES IS INDEXED IN SCOPUS

Giovanna Fossa • Brent D. Ryan •
Carmelo Ignaccolo
Editors

Small-Town Renaissance

Bridging Technology, Heritage and Planning in Shrinking Italy

Springer

Editors
Giovanna Fossa
Department of Architecture, Built Environment and Construction Engineering
Politecnico di Milano
Milano, Italy

Brent D. Ryan
Department of Urban Studies and Planning
Massachusetts Institute of Technology
Cambridge, MA, USA

Carmelo Ignaccolo
E. J. Bloustein School of Planning and Public Policy
Rutgers University
New Brunswick, NJ, USA

ISSN 2198-7300 ISSN 2198-7319 (electronic)
Research for Development
ISBN 978-3-031-89785-6 ISBN 978-3-031-89786-3 (eBook)
https://doi.org/10.1007/978-3-031-89786-3

This Springer imprint is published by the registered company Springer Nature Switzerland AG
The registered company address is: Gewerbestrasse 11, 6330 Cham, Switzerland

Acknowledgments

The authors thank Fondazione Rocca, MIT Italy (Serenella Sferza), and Fondazione Politecnico di Milano (Daniele Bignami) for their support in making this research project and book possible.

Contents

About the Editors

Giovanna Fossa is Full Professor of Urban Design and Planning, Department of Architecture, Built Environment and Construction Engineering, Politecnico di Milano. She teaches Regional Planning and Regenerative Urban Design at the School of Civil and Environmental Engineering. She has gathered a long international experience (teaching, research and projects) in the fields of urban and territorial regeneration, regional and landscape planning, tourism and production, place making and branding. A special focus is always on cross-disciplinary and cross-border themes. Her collaboration networks include MIT Department of Urban Studies and Planning; Yale School of Architecture; SUPSI Department of Environment, Constructions and Design; Regional Plan Association New York; Fundacion Metropoli Madrid. She is member of the Collegio di Milano and ULI Italy Scientific Committees. She is author/co-author of several publications, among them *Hybrid Factory Hybrid City* (Actar 2022), *Shaping Spaces for ever-changing mobility* (TeMA 2020), *The Design of Urban Manufacturing* (Routledge 2020), *Planning Talks* (Maggioli 2018), *ITATOUR* (F. Angeli 2012).

Brent D. Ryan is Vice Provost and Professor of Urban Design and Public Policy in MIT's Department of Urban Studies and Planning. His research focuses on the aesthetics and policies of contemporary urban design, particularly with respect to current and pressing issues like deindustrialization and climate change. Professor Ryan has published in numerous journals including the *Journal of Urban Design*, *Urban Design International*, *Journal of Planning Education and Research*, and the *Journal of the American Planning Association*, which awarded his article *Reading Through A Plan* its best article of 2011. Professor Ryan is co-editor in chief of the Journal of Planning History, published by Sage. He is currently conducting design and planning studies of shrinking cities, rural settlements, and regions in Italy, Japan, and India, and his new book project examines the intersection of tourism, mapping, and travel guides during the 19th and early-to-mid 20th centuries. Professor Ryan holds a B.S. in Biology from Yale University, a M. Arch. from Columbia University, and a Ph.D. in Urban Design and Planning from MIT.

Carmelo Ignaccolo is Assistant Professor of Urban Design, Technology & Climate at Rutgers University's E. J. Bloustein School of Planning and Public Policy. Professor Ignaccolo's work bridges urban design and city planning scholarship through spatial analytics, community-based research, and historical methods. His scholarly research focuses on spatial inequality, environmental vulnerabilities, heritage challenges, and multisensory urbanism. He has published articles in *Cities*, *Magazine on Urbanism*, and the *Journal of Urban Technology*. His digital mapping work has been featured in *The New York Times* and *Bloomberg CityLab* and exhibited at Architecture Biennials in Seoul (2019), Shenzhen (2020), and Venice (2023). Before joining Rutgers, Carmelo held research and teaching positions at MIT DUSP and Columbia University GSAPP. Beyond academia, he practiced as an Urban Designer at AECOM in New York City and as an Urban Planner for the United Nations Habitat Programme in Nairobi. He received his Ph.D. in Urban Planning, Design, and Technology from MIT DUSP, his Master's in Architecture and Urban Design from Columbia University GSAPP, and a combined Master's and Bachelor's degree in Architecture-Engineering from the University of Catania, Italy.

Small Towns, Big Visions: Technology and Heritage in Small-Town Planning

Carmelo Ignaccolo and Brent D. Ryan

Demographic Decline in the Digital Post-COVID Age

Demographic shrinkage is rapidly becoming a global phenomenon. By the late 2070s, the global population aged 65 and older is projected to reach 2.2 billion, surpassing the number of children under 18 (United Nations Department of Economic and Social Affairs, Population Division, 2024). Ignoring the historical causes, ongoing effects, and lasting implications of this era of demographic decline risks overlooking one of the most disruptive changes in human history.

Initially concentrated in Western countries and aging Asian nations such as South Korea and Japan, this mid-to-late-twentieth-century trend is now spreading to developing regions (Ryan, 2020). According to demographers, nations that fueled significant demographic growth in the latter half of the twentieth century and the early decades of the twenty-first—such as India, China, and much of Latin America—are poised for a shift. In other words, humanity is on the brink of a new global epoch: *the age of depopulation*. Political economist Nicholas Eberstadt predicts that, for the first time since the Black Death of the fourteenth century, the global population will decline. Yet this new cycle of depopulation will differ profoundly from its historical precedent (Eberstadt, 2024). While a catastrophic epidemic drove the Black Death, the coming decline will arise largely from human

C. Ignaccolo (✉)
E. J. Bloustein School of Planning and Public Policy, Rutgers University, New Brunswick, NJ, USA
e-mail: carmelo.i@rutgers.edu

B. D. Ryan
Department of Urban Studies and Planning, Massachusetts Institute of Technology, Cambridge, MA, USA
e-mail: bdr@mit.edu

G. Fossa et al. (eds.), *Small-Town Renaissance*, Research for Development,
https://doi.org/10.1007/978-3-031-89786-3_1

choices—or, perhaps, from profound societal transformations and ineffective policies.

Depopulation, however, is not happening as a stand-alone phenomenon. It is unfolding during a historical moment marked by extreme digital innovation. The challenges and opportunities brought by digitalization to society are vast, yet their impact is unevenly distributed. Over the past decades, large metropolitan areas have been outfitted with advanced digital infrastructures, including sensors, city dashboards, and fiber-optic internet. Cities, as hubs of wealth, talent, and job opportunities, have reaped the benefits of these advancements (Florida, 2008; Sassen, 1991). In contrast, rural towns have frequently been left behind, lacking comparable access to digital tools and resources. Despite the interconnectedness of global capital flows and information within the framework of planetary urbanization, the spatial marginalization of rural areas persists. This growing divide highlights the uneven geography of digital transformation, raising critical questions about inclusivity and equity in *the digital era of demographic decline*.

Existing literature on demographic decline has largely followed two key trajectories. First, the focus on economic development and sociological analysis has often overlooked the built environment, which serves as both a contributing factor to and a consequence of demographic shrinkage. Second, restoration efforts have predominantly targeted the physical environment as the primary site of intervention, often attributing to it a near-maieutic power to regenerate both physical structures and the social fabric of small towns. Furthermore, both approaches have leveraged technological innovation primarily as an analytical toolkit—using digital tools to document existing built fabric and to visualize socio-demographic transitions. This methodological reliance has left significant gaps in addressing the interplay between digital tools, built heritage, and the socio-economic transformations of shrinking communities. More precisely, there is still little explored on how creative digital tools can move beyond mere monitoring to drive a new system of public policy interventions, foster entrepreneurship, and support new forms of living and working in non-metropolitan settings (Fig. 1).

In other words, through this edited volume, we investigate the following research questions: how can digitization create new development pathways while leveraging the heritage and local production of small towns? What transformative actions are required locally in terms of physical planning and public policy to support these changes? Finally, as small towns evolve and reimagine their identity, how can local communities actively participate in the digital innovation process?

To address these questions and methodological shortcomings, this edited volume seeks to bridge these divergent approaches. It uses the lens of digital technologies to offer empirical analyses, propose design-oriented solutions, and connect centuries-old built heritage to twenty-first-century trends, particularly shifts in the geographies of work. It does so by innovating the current literature and providing a fresh post-COVID look at what is happening in rural areas. Over the past four years since the pandemic hit, it has become clear that while a broad repopulation trend may be unrealistic, there are signs of localized "renaissances" and a renewed awareness of what it means to live in non-metropolitan areas. For example, the interior design

Fig. 1 Data collection of 360° images during the "Digital Paths" project, led by the NGO Liminal in 2023, in Guadagnolo, Monti Prenestini (Lazio, Italy). Participatory mapping using high-resolution imagery helps bring the unmapped narrow streets of small towns into the digital realm. Source: Liminal (2023)

market has experienced a surge in demand for home office spaces, alongside rising real estate values in areas outside traditional urban centers that maintain good access to public transport. Some thinkers have even described this trend as a new "suburban wave". These shifts reflect a growing desire for lifestyles that combine professional flexibility with enhanced access to quality living conditions beyond the constraints of traditional urban cores.

As demographic decline and digital connectivity emerge as central dynamics shaping the coming centuries, humanity faces an urgent need to develop innovative urban planning frameworks and actionable strategies tailored to an increasingly diverse array of physical environments. Whether addressing the densification of metropolitan regions or the revitalization of small towns and rural settlements, these strategies must balance the demands of digital transformation with the celebration of local identities and cultural heritage.

This book, titled *Small-town Renaissance*, encapsulates a reflection on these transformative trends, drawing inspiration from the term 'renaissance' to frame its approach. We deliberately chose this term for three reasons. First, historically, the term 'renaissance' symbolizes a period of rebirth and renewal, aligning with the challenges and opportunities of reimagining life in shrinking yet technology-equipped settlements. Second, the word renaissance emphasizes transformation through building on what already exists—a powerful metaphor for adapting long-standing settlements and infrastructure to meet contemporary needs. Third, like its Florentine counterpart, a small-town renaissance focuses on centering humankind, celebrating creativity and adaptability in shaping the built environment.

Fig. 2 Fieldwork activities in Centuripe (Enna, Sicily) with MIT students, local authorities, and faculty guests from the University of Catania. Source: photo by Ryan (2020)

Given the global scope of demographic and spatial transformations, this introduction chapter provides an up-to-date overview of recent trends in shrinking studies across diverse contexts, including Italy, Switzerland, Japan, and China. These regions offer valuable insights into how shrinking populations intersect with local cultural, economic, and technological dynamics, and they highlight the importance of place-specific strategies in addressing global challenges. Italy is the testbed site of this edited volume, drawing from on-the-ground experience conducted in Italy by MIT and Politecnico di Milano research teams in 2022 and 2023 (Figs. 2 and 3).[1]

Shrinkage, Heritage, and Technology in the Italian Countryside

Within the context of this research at the intersection of demographic shrinkage, heritage, and technology, Italy plays a unique role: it is aging (24.3% of the Italian population was aged 65 years and older as of January 2024, according to ISTAT, 2023), it is rich in heritage (with 60 listed UNESCO heritage sites, Italy is the state party with the most World Heritage Sites), and it has been pushing an unprecedented coverage of digital infrastructure (Italy ranks in third place out of 28 countries, in

[1]An MIT website dedicated to the research project is available at the following link: https://sites.mit.edu/smartvillagesofitaly/.

Fig. 3 Fieldwork activities in Valtellina (Lombardy) with MIT and Politecnico di Milano research teams. Source: photo by Ryan (2020)

the European FTTH/B coverage ranking and second in terms of annual growth rate in 2020 according to Open Fiber, 2021).

In Italy, the challenge of integrating small settlements with advanced technologies is not new. In 1989, architect and educator Giancarlo De Carlo spearheaded the redevelopment of Colletta, a village in the township of Castelbianco (Liguria). The village, abandoned for years, had a population of just five people in 1971. It was eventually purchased in its entirety by a developer from Alessandria, a city in Piedmont, with the vision of transforming it into a haven for "white eagles" (Fig. 4).

As De Carlo explains, "white eagles" are individuals who work independently, seek isolation, and rely on computers and information technology to stay connected to the world. He described them as people who are "not only in touch but may even direct operations in other, more crowded places" (De Carlo, 2004). Despite being an urbanite himself, De Carlo found Colletta's urban fabric and its sequence of open and enclosed public spaces immensely appealing.

In his writings, De Carlo noted that modern technology, such as wires and cables, could be discreetly concealed behind plaster without causing significant alterations to the village's historic structures. Today, Colletta continues to brand itself as a *Borgo Telematico* ("Telematic Village"), honoring its legacy as a pioneering example of what many small towns now aspire to achieve in the post-COVID era. With the advent of Fixed Wireless Access Networks, which reduce reliance on physical cable infrastructure, Colletta's early vision has become even more relevant.

The case of Colletta, however, does not illuminate us on a specific front: it was a fully abandoned town with no residents seeking to shape a new future. While the architectural intervention is cutting-edge, the absence of an active community in

Fig. 4 The village of Colletta in Castelbianco, Liguria, underwent a significant renovation in the early 1990s. Led by architect and educator Giancarlo De Carlo, the project transformed the site into a pioneering "Telematic Village". Source: Wikimedia Commons

this transition rendered the project more of an architectural exercise at the village scale. It required little engagement with public policy or community involvement in integrating technologies. This lack of a participatory dimension highlights a critical gap in understanding how technological innovation can be meaningfully woven into the fabric of small towns where residents are still present, grappling with demographic shrinkage, and seeking sustainable futures.

A Policy Overview of Italy's Planning Efforts in Small Towns

Since World War II, the Italian population has experienced relevant changes in terms of demographic distribution along the Italian peninsula (Golini et al., 2003). For example, intra-national migration patterns went from the South to the Center and the Northwest, from the East to the West, and, generally, from small and medium urban centers to larger cities, from the mountains to the hills and plains, and from rural to industrial areas (Birindelli & Balletta, 1976). Since the middle of the 1970s, the level of migration has reduced its magnitude and migration from the South to the North has also decreased. Nevertheless, migration from rural to urban areas has continued to be relevant, maintaining a slow yet constant trend. Intra-national migration contributed to shaping economic differences between Italian regions, and it has also been responsible for the depopulation of some areas. On the one hand, migration has depopulated some areas, but on the other hand, it has populated other regions. So, at the same time, it has contributed to both marginalizing some areas and making other territories more developed (Morettini & Chiapparino, 2018).

In Italy, as in the rest of Europe, what began as a process of post-war migration to richer urban areas—as well as international cities—has turned into a chronic problem. As a result of weak local markets, public services became underutilized, poorly maintained, and often unviable to the point of being withdrawn. Additionally, local living conditions and quality of life deteriorate, unemployment rises, and skilled labor becomes scarce, causing the emergence of abandonment and obsolescence. This further erodes the attractiveness of small towns, and it enables the rotation of a downward spiral of demographic decline through falling fertility rates and enforced aging of the remaining population. Indeed, continued depopulation is predicted to become the new normal in many peripheral rural regions over the coming decades; therefore, there is an urgent need to deploy local policies to address questions of rural governance and territorial inequalities between cities and rural towns (Pires de Almeida, 2017).

Thanks to a hybrid policy response—positioned in between a reverse in declining demographic trends and a spatially differentiated approach—Italy started developing in 2012 a National Strategy for its Inner Areas, named SNAI (Barca & Lucatelli, 2014), a national policy named in the 2014 "Programma Nazionale di Riforma"—PNR (tr. national reform program) and defined in the 2014–2020 "Accordo di Partenariato" (tr. partnership agreement), also confirmed in the 2021–2027 cycle.

SNAI's goals were to create developmental tools that could potentially foster a series of improvements in the well-being of people living in the most remote and depopulated areas of the peninsula (Pezzi & Urso, 2016). From a spatial classification standpoint, Italy's so-called 'inner areas' occupy 60% of the country, 52% of the municipalities with only 22% of the population. From a methods point of view, SNAI works as a territorial co-planning operation carried out by the Agency of Territorial Cohesion under the supervision of the Ministry of Economic Development and coordinated by a technical committee based in Rome with the task of assisting local actors (Fig. 5).

While SNAI has often been recognized as a ground-breaking and effective policy scheme that successfully connected the needs of local communities to central government action plans, the recent COVID-19-induced digital transformation (e.g., remote working, online learning, etc.) requires new development frameworks. These frameworks could link enhanced digital access in remote areas to their exquisite built heritage features and economic activities (often centered on the agriculture sector). Surprisingly, regions with consolidated depopulation trends (such as Sicily, Basilicata, Friuli Venezia Giulia, Abruzzo, and Umbria) are now positioned at the frontline of a "Smart Villages" revolution. In fact, the Open Fiber map on digital

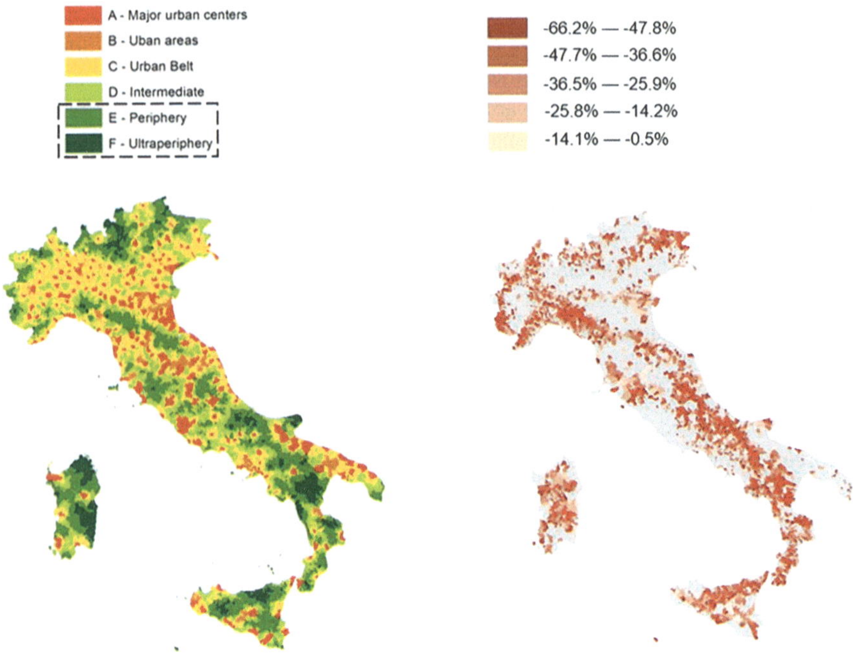

Fig. 5 (left) Map showing the Agency for Territorial Cohesion's accessibility rankings for each Italian municipality (2014); (right) map illustrating demographic decline rates across Italian municipalities through ISTAT data from 2011. Source: image by Ignaccolo (2020)

access in Italy shows that these regions have the highest % of deployed optic fiber networks (Banda Ultralarga, 2024).

Situating Italy's Trends and Policies in a Shrinking World

Italy's National Strategy for Inner Areas (SNAI), initiated over a decade ago, is now being bolstered by post-COVID European funds as part of the National Plan for Rebooting and Resilience (Piano Nazionale di Ripresa e Resilienza or PNRR). Similar policy approaches are emerging in other European countries, such as Spain, Greece, Finland, and Sweden, as well as in the Far East. Broadly speaking, the digital transition accelerated by the COVID-19 pandemic has created a historic opportunity to rethink the future of shrinking villages around the world.

In Italy, unprecedented public investments are planned over the coming years to upgrade digital infrastructure in inner areas as part of the PNRR. These efforts build upon the groundwork laid by the Ultra-Broadband National Plan (BUL) in 2015, which targeted market-failure areas. A key emerging technology in this context is the combination of the Fixed Wireless Access (FWA) paradigm with 5G. FWA has the potential to provide long-range Internet access to inland and remote areas, overcoming the challenges posed by difficult terrain that have historically hindered the rollout of fixed cable infrastructure.

This scenario presents significant opportunities for the conservation and promotion of cultural heritage. Digital technologies offer innovative tools for documenting and preserving historical buildings and enable the enhanced appreciation and utilization of the cultural heritage that defines inner areas. Augmented reality, digital platforms, and mobile applications, among other tools, can greatly enrich engagement with these assets. Supporting this potential, the PNRR has allocated 600 million euros for the conservation and valorization of rural architecture and over 1 billion euros for the enhancement of hamlets and villages.

Given these specific historical circumstances, this section reviews a series of policies and approaches taking place outside Italy to help situate the discussion within the broader contexts of Europe, Switzerland, Japan, and China.

Europe: A Transnational "Rural Vision"

At the European level, rural areas cover 83% of the EU territory and host 30% of the EU population with approximately 137 million people (Eurostat, 2023).

In 2021, the European Commission proposed "A Long-Term Vision for the EU's Rural Areas" aimed at providing a "*long-term vision from, by and for stronger, connected, resilient and prosperous rural areas by 2040*" (European Commission, 2021). Having recognized the importance of issues such as population decline and aging in rural areas, the document "*aims to address these challenges and concerns, by building on the emerging opportunities of the EU's green and digital transitions and on the lessons learnt from the COVID-19 pandemic, and by identifying means to improve rural quality of life [...]*", also acknowledging rural areas as "*active players in the EU's green and digital transitions*".

In March 2024, the European Commission showed the progress achieved on the long-term vision for rural areas in the report "The Long-Term Vision for The EU's Rural Areas: Key Achievements and Ways Forward" (European Commission, 2024c). Among the achievements, the EU included the introduction of an EU Rural Observatory for the improvement of rural data collection and dissemination, the creation of a "rural revitalization platform" (launched in June 2023) allowing common access to information and resources targeting rural communities and stakeholders, and the EU rural toolkit (launched in February 2024), which acts as an interactive guide collecting the available EU funding opportunities for rural areas for easier consultation and access (European Commission, 2024b). In the context of the "Rural Vision," in the period 2021–2024, a €253 million budget was devolved to start 60 rural research and innovation projects (European Vision, 2024). With dedicated actions, the "Rural Vision" also supports the development of smart villages and rural connectivity, mobility, and tourism (European Commission, 2024a).

According to the European Commission, 43% of land in rural areas is used for agriculture (European Commission, 2024c). Developing a sustainable bioeconomy in rural areas is regarded as a trend that can find innovative opportunities to address circularity and decarbonization. Despite this, "*Over time, the contribution of agriculture, forestry, and fisheries to rural regions has decreased both in economic and employment terms to 12% of all jobs and 4% of gross value added, while crucially maintaining food security in the EU*" (European Commission, 2021).

Finally, already experiencing the extreme impacts of climate change, rural areas could provide diversified solutions for climate change mitigation and circular economy (Atkinson & Atkinson, 2023). In general, they will then play a central role in the framework of the "European Green Deal" as potential hubs of sustainable activities.

Switzerland: Fiber Infrastructure Across the Alpine Landscape

Rural areas comprise a significant part of the Swiss landscape (UFAM, 2024). According to the 2020 Geographic Levels Statistics, 575 Swiss municipalities are classified as rural, representing 49% of Swiss municipalities. These municipalities host 14% of the population and 9% of the jobs. They cover a large area, equivalent to 57%, considerably more than the 17% urban municipalities occupy. However, urban municipalities are home to 65% of the population and 76% of the jobs (UST, 2024).

Similar to EU Member States, Swiss rural areas face several challenges, including complex topography, reduced accessibility, and, as mentioned above, a low population and business density, which results in economic disadvantages (SECO, 2024a).

At the national level, the Confederation has developed a strategy called the "Swiss Territorial Project," which provides a framework for guiding and supporting decisions on the future territorial development of the country. The aim of this strategy is to promote cooperation between the various levels of authority for organised urbanization and infrastructure development in order to reduce the impact and use of land (Ufficio Federale dello Sviluppo Territoriale ARE, 2024). This provides the framework for coherent territorial development (SECO, 2024b).

Regional development projects specifically promote agriculture and cooperation between different areas. These projects serve as promotional tools, supporting rural development ideas to create added value in the agricultural sector by maximizing the potential of the region (UFAG, 2024). There are currently 27 regional development projects in Switzerland—visible on the Swiss mapping portal www.map.geo.admin.ch of the Confédération Suisse (2024)—10 of which are located in the Alpine region. A notable example is the project in Val Poschiavo, which aims to promote organic production and the processing of local products.

Switzerland is one of the most advanced countries in internet access, with broadband coverage of 99.8% nationwide and 98.6% in rural areas, well above the

European Union average of 89.7% (Fig. 6). The fibre network covers 67.5% of Swiss rural households, compared to 20.1% in the European Union. The urban-rural gap is, therefore, smaller in Switzerland than in other countries (Bürgin, 2021).

At the national level, the Federal Council has developed the "Gigabit Strategy" with the aim of providing high-speed and high-performance Internet access

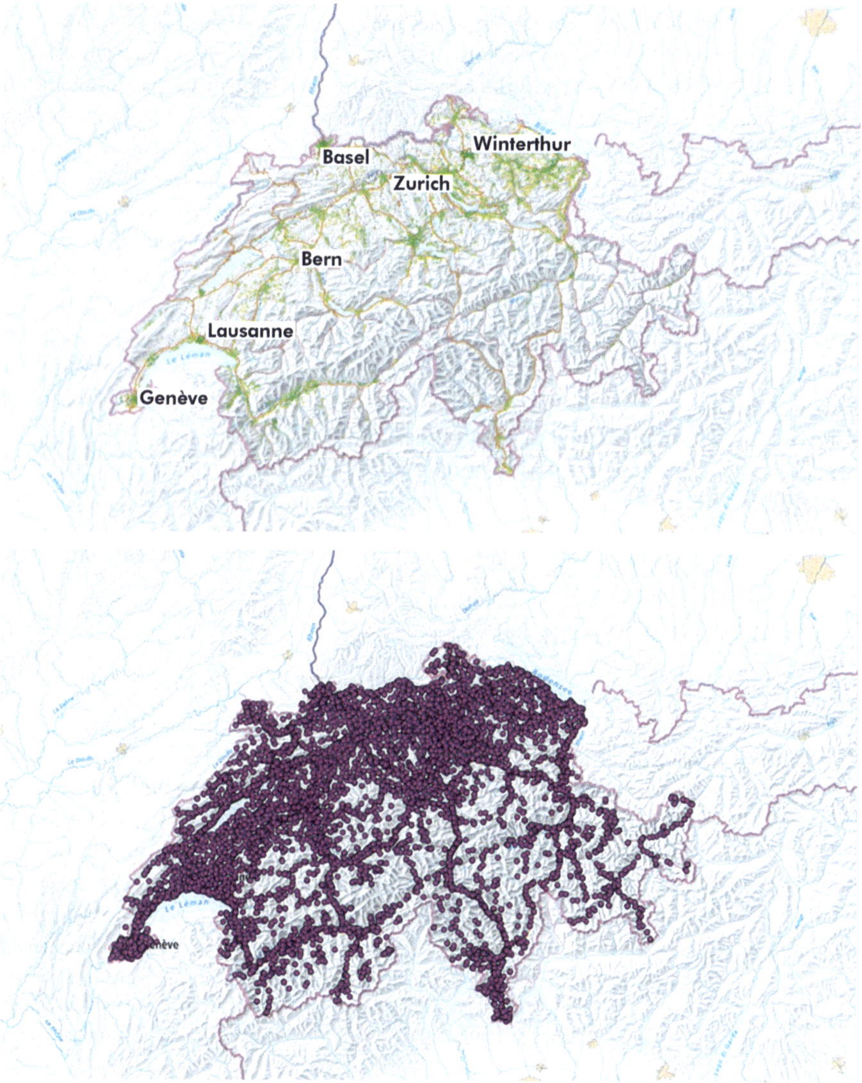

Fig. 6 (top) Internet availability via the fixed network—Download ≥300 and ≥500 Mbit/s (Federal Office of Communications OFCOM); (bottom) Mobile phone base stations 05.12.2024 (Federal Office of Communications OFCOM). Source: Cartes de la Suisse—Confédération Suisse (2024) https://map.geo.admin.ch/. Retrieved December 6, 2024

throughout the country. The aim of this initiative is to take full advantage of digitization in Switzerland and maintain the attractiveness of more remote areas. The strategy includes a support program to expand the network outside urban centres, encouraging private companies to invest in areas where it is not economically advantageous to develop telecommunications infrastructure, thus bridging the cost-benefit gap. The support is primarily focused on extending fibre-optic connections to households and less populated areas (Consiglio Federale, 2023).

Thanks to digitalization, on which mountain regions have high hopes, it is possible to attract new types of workers that do not necessarily rely on agriculture or tourism. In fact, the mountain regions are focusing on proposing new applications for tourists, the development of digital platforms, and the extension of fiber-optic networks. Many villages in the Alps are opening coworking spaces to attract professionals to remote valleys (Bondolfi et al., 2019). For example, in the Hérens Valley in the canton of Valais, a couple of former travellers decided to open a co-living space in the small village of Haudères after various co-living experiences around the world. Their aim was to welcome digital nomads and involve them in local life by offering activities that revitalize the village (Killé Elsig, 2023; Police, 2023). Another example is the "miaEngadina" project in the Inn Valley (Engadine), which provides a co-working space and information on accommodation in the region, with the possibility of organizing activities such as excursions and tastings (Bondolfi et al., 2019). Even in the canton of Ticino, valleys are structuring development opportunities in this direction. An important example is the Blenio Valley, which has been offering a coworking space for about a year. This space is complemented by training courses and community events and offers meeting rooms and workstations for different needs (Cowo, 2024).

Japan: AI and Digital Hubs amid Demographic Decline

As of 2023, Japan's total population is 124.35 million, with 36.23 million people aged 65 and older accounting for 29.1% of the population (Cabinet Office, 2024). By 2037, the Japanese government predicts that 33.3%, or one out of every three citizens, will be 65 or older (Cabinet Office, 2024). In light of these trends, how are the Japanese national and municipal governments using technologies and the promotion of remote work, among other measures, to address Japan's declining birthrate and aging population?

The use of technology is an important policy goal of the Japanese government. In fact, in September 2021, the Japanese national government created the Digital Agency to realize "*the happiness of each and every individual through the use of digital technology*". And every year since 2021, the Cabinet has approved a priority plan for the advancement of a digital society (Digital Agency, 2024). In addition, one of the priority issues identified by the Digital Agency is the labor force shortage due to population decline and the concentration of population in metropolitan areas, which will result in the inability to maintain necessary public services. Also, one of the Digital Agency's initiatives is to develop and operate a digital infrastructure that can be used commonly across municipalities, replacing the current public service systems maintained by the national government and local governments on their own (there are approximately 1800 local governments in Japan). In this way, the Digital Agency aims to maintain and strengthen public services in a society with a declining population through digital technology. The above trends in the national government are also true in local governments, where various advanced efforts have been made. For example, Osaka Prefecture has an ICT introduction support project subsidy mechanism for introducing nursing care robots and using ICT to improve productivity and alleviate staffing shortages at nursing care facilities (Osaka Pref, 2024).

Like Western countries, the percentage of remote workers in Japan increased after COVID-19. According to a national survey, the remote working rate in 2021 was approximately 27%, and the percentage of remote workers has been increasing in Tokyo and regional metropolitan areas (the survey sample size is approximately 40,000 people). Interestingly, less than 25% of respondents did not consider commuting time or distance when choosing where to relocate. Furthermore, the most common reason for moving was to live in a larger home, about 40% (Ministry of Land, Infrastructure, Transport and Tourism, 2022). Given this situation, there is an important opportunity to attract more people from urban areas to rural areas.

In the midst of this trend, local governments have been making progress in establishing remote work centers in rural areas. For example, Kyoto City established a regional remote work center utilizing the site of a closed elementary school located in the mountainous area of the Keihoku (Ministry of Internal Affairs and Communications, Regional Communications Promotion Division, Information Distribution Administration Bureau, Regional Informatization Planning Office, Local Administration Bureau, 2022). The former elementary school site now has "a remote work area" and "a creative area." A remote work environment was developed through these areas to encourage co-creation and collaboration among local residents, universities, and businesses. In the creative area, there is also a space for wood processing and cooking classes, making use of the abundant wood resources in the area and the former arts and crafts and home economics rooms of the elementary school (Kyoto Satoyama SDGs Lab, 2021; Ministry of Internal Affairs and Communications, Regional Communications Promotion Division, Information Distribution Administration Bureau, Regional Informatization Planning Office, Local Administration Bureau, 2022).

Furthermore, local governments are also using local e-currencies for regional development. This is because local e-currencies can stimulate local consumption. For example, Nishi-Izu Town in Shizuoka Prefecture, well-known for its pristine coastline, hot springs, and sunset views, introduced its own local currency, the "Sunset Coin," in 2020, despite having a population of less than 10,000 (Nishi-Izu Town Tourist Association, 2024; Ministry of Internal Affairs and Communications, Regional Communications Promotion Division, Information Distribution Administration Bureau, Regional Informatization Planning Office, Local Administration Bureau, 2024). This currency is introduced based on a local currency platform operated by a private company (Fig. 7). There are several companies

Fig. 7 Coastline of Nishi-Izu Town, where the local government introduced in 2020 the e-currency named "Sunset Coin". Source: Batholith, Wikimedia Commons, 2012

that operate local currency services, including the use of blockchain technology and the ability to fluctuate the value of coins during events (Mitsubishi Research Institute, Inc, 2024).

Finally, various efforts are being made in each region to promote tourism. For example, the Niseko area in Hokkaido, a world-famous ski resort with a population of less than 30,000, has many non-Japanese tourists. For this reason, three towns in the area are collaborating to operate a multilingual translation AI chatbot. This project has contributed to the revision of marketing strategies based on analysis of trends in the information people have viewed, the active dissemination of information by local businesses, and the expansion of consumption in the region. The project utilizes about 50% of Hokkaido's grant for regional development (Ministry of Internal Affairs and Communications, Regional Communications Promotion Division, Information Distribution Administration Bureau, Regional Informatization Planning Office, Local Administration Bureau, 2024).

China: Confronting the Challenges of a Looming Shrinking Future

In the context of large-scale urbanization in China over the past decade, urban shrinkage has largely been overlooked in mainstream research on urban growth. This neglect is understandable, given China's ongoing rapid urbanization and the emphasis of urban planning on expansion (Deng et al., 2019). However, recent studies suggest that urban shrinkage is becoming increasingly apparent and could pose a significant challenge to future urban development (Liu et al., 2020). By comparing the 2010 and 2020 Chinese censuses, a study has shown evidence that 180 out of 653 cities were identified as shrinking cities, including 140 county-level cities (Meng & Long, 2022).

On March 31, 2019, the National Development and Reform Commission issued the "Notice on Key Tasks of New Urbanization Construction in 2019", formally introducing the concept of "shrinking cities." Despite this, the widespread occurrence of urban shrinkage across various economic regions in China has not been met with sufficient attention or understanding of the underlying causes and mechanisms driving population decline (Yang et al., 2021).

The majority of shrinking cities are small to medium-sized settlements and are distributed across the central (Meng & Long, 2022), southeastern (Hu et al., 2021), western (Li et al., 2024), and northeastern regions (Hu et al., 2021; Li et al., 2024; Meng & Long, 2022), with the northeastern region being particularly affected (Yang et al., 2021). The three northeastern provinces (Liaoning, Jilin, and Heilongjiang), known as China's "Rust Belt" (Fig. 8), were once heavy industrial powerhouses, with numerous state-owned enterprises, especially in the fields of iron and steel, machinery manufacturing, petrochemicals, and automobile manufacturing (Li, 2004). The three provinces have recently experienced the lowest birth rate in the country due to the vigorous and strict implementation of family planning policies (Yang et al., 2021), which has led to a decline in the population and a loss of urban population.

Fig. 8 Photo of Harbin's city center with St. Sophia Cathedral in the foreground. Harbin is the capital of the rapidly shrinking Heilongjiang Province. Source: Joann Pittman (License: CC BY NC ND 2.0)

From an economic standpoint, as China transitions to a market economy, intensified international market competition poses significant challenges to these long-time-operating industries. The government has attempted to restructure the industrial heritage of these cities by promoting high-tech manufacturing and modern service industries. Nonetheless, the difficulties and resistance encountered during the transition have led to economic contraction and depopulation in some cities (He et al., 2017), such as the former industrial cities of Jixi and Hegang in the northeast (Deng et al., 2019).

In addition to economic restructuring and industrial decline, uneven economic development has led to a concentration of resources and talent in a few developed regions, exacerbating urban shrinkage in some areas, especially in the vicinity of large metro areas; the so-called 'siphon effect' (Chen et al., 2024; Deng et al., 2019). Shrinkage occurs not only in economically underdeveloped small cities but also in relatively developed regions, such as the Pearl River Delta and the Yangtze River Delta, dominated by export-oriented processing industries. With the global economic crisis leading to a reduction in international trade orders and rising labor costs, the manufacturing sector in these regions is shifting to other countries in Southeast Asia.

Policy implications and spatial administrative adjustments have also contributed to urban contraction and deidentification strategies within urban areas. Some examples of urban shrinkage are the result of government interventions through planned resettlement projects, such as the inner-city de-population of the mega-cities of Beijing, Shanghai, and Tianjin, which occurred through inner-city regeneration programs (Li & Mykhnenko, 2018; Yang & Ley, 2019). The market value of land in city centers is the main reason for inner-city regeneration schemes.

The causes of China's urban shrinkage remain controversial (Long & Gao, 2019b), however, "hollowing-out of the labor force," "outflow of population," and "degradation of resources" are widely acknowledged in most studies as common factors contributing to the phenomenon of urban shrinkage (Long & Gao, 2019a, p. 23). In general, "expansion within shrinking" is the notable characteristic of urban shrinkage in China (Hu et al., 2021; Zhou et al., 2019, p. 25), whereby population declines and land expansion occurs simultaneously. Several key factors drive this phenomenon. Local governments have made urban expansion and increased fiscal revenues their primary goals. Despite the population loss, many cities continue to expand their urban areas to increase land use and attract investment. In addition, reliance on revenue from land sales encourages local governments to expand urban land to generate revenue without adjusting for actual demographic changes. China's long-standing pattern of economic growth, which relies heavily on large-scale infrastructure investment, has tended to allocate development zones and construction projects to cities with relatively shrinking economies, aiming to promote regional balance and balanced economic development, even though this may lead to land expansion rather than population growth. These factors lead to the apparent paradox that even though the population is declining, the built-up areas of these cities are still expanding (Long & Gao, 2019b, p. 12).

Book Structure and Chapters' Contributions

Small-Town Renaissance is organized into two main areas of scholarly investigation. Within these two groups, authors have employed a variety of research methods, including big data analytics, on-site fieldwork, community engagement, and pedagogical endeavors at MIT and Politecnico di Milano. As co-editors, we have chosen to structure the chapters to highlight the multifaceted roles that digital tools and technologies play in shrinking areas. Instead of organizing the chapters by scale (e.g. regional, township, neighborhood), we have opted for a spatially cross-cutting approach that centers on the notion of technology, using digital tools as guiding lenses through which to observe, learn and implement development strategies in small towns.

Part I focuses on the role of digital technologies as tools for heritage management and enhancement. This section contains four contributions that respectively examine: the application of digital tools in monitoring heritage sites in small towns; the uncovering of rich cultural heritage assets (e.g., creative practices, artisans) through a digital platform; the mapping of abandonment through data-driven participatory practices; and the envisioning of a diffused sport-heritage landscape.

Part II explores the digital dimension from a socio-economic perspective, reflecting on both enduring and temporary changes in the post-pandemic conditions of small towns. More precisely, this section examines the role of remote work as a development strategy for certain small towns, considering socio-economic factors, infrastructural networks, and unique aspects of the built environment. The contributors provide insights, analyses, and strategic recommendations on the post-pandemic paradigm shift in the spatiality of remote work across Southern Europe (e.g., southworking), the use of regional planning schemes to advance remote work development strategies (e.g., Valtellina), and the deployment of ubiquitous technologies ranging from sensors and highspeed internet access in outdoor settings to electric mobility on mountain trails (e.g., Alpine rural tourism).

The two sections are complemented by a final chapter that gathers insights from stakeholders and experts involved in the research on the ground. It features voices from local communities alongside bold scenarios and visions with national and international perspectives, addressing the challenge of revitalizing historically anthropized territories now at risk of abandonment.

By anchoring our analysis in the concept of renaissance, this volume invites readers to consider how a forward-looking synthesis of technology, design, and policy can foster vibrant, sustainable futures for communities of all sizes.

Part I: Heritage Management and Enhancement Through Digital Tools

Italy's heritage assets are often celebrated for their unparalleled diversity across regional areas and millennia-old stratifications. However, high maintenance costs and intense competition among various sites frequently hinder renovation efforts and strain the financial resources generated through ticket sales. In this first chapter, Boniotti and Pimpinella—a heritage conservationist and a telecommunications engineer, respectively—join forces to demonstrate how mobile radio data extracted from cellphone towers can aid heritage monitoring by assessing visitor numbers and the popularity of specific areas. Drawing on a dataset of Internet traffic, the authors offer empirical evidence to illuminate post-pandemic repopulation trends that have often been discussed anecdotally. Their scholarly inquiry focuses on the province of Sondrio (Lombardy), where they explore the density of heritage site users. Specifically, the authors investigate whether mobile network data can be utilized to evaluate the potential for cultural heritage engagement and how this potential has evolved in the Valtellina area following the pandemic. This thought-provoking and interdisciplinary work presents a detailed analysis of both funding schemes and infrastructural upgrades in the Valtellina area, covering both the heritage and telecommunications sectors. Boniotti and Pimpinella discover a post-pandemic increase in mobile users within Sondrio Province, with an even more significant rise (40%) near heritage sites. Notably, the most impactful finding of their research is the identification of an increasing trend in mobile user activity during weekdays, suggesting a new form of residential life in the valley that extends beyond traditional weekend visits.

While Boniotti and Pimpinella employed a data-driven, top-down analysis of demographic trends, Goh and Halim masterfully combine the granularity of Internet traffic data with a comprehensive, people-oriented, and place-sensitive investigation of the network of small and medium enterprises (SMEs). Drawing from a six-month-long community engagement with stakeholders in Lombardy and Sicily, Goh and Halim bring to life the 'territorial capital' of small businesses spread across the towns of Vizzini (Catania), Centuripe (Enna), and Sondrio. Their final outcome is the ideation of an innovative e-commerce platform that highlights the stories behind exceptional small businesses, ranging from local farmers and Greek-vase style artisans to saffron cultivators. Goh and Halim illustrate the potential of e-commerce platforms to make local products accessible to distant customers while simultaneously unearthing and celebrating the stories of the people behind these products. In their design effort, the authors introduce a compelling, interactive, place-based storytelling approach that incorporates the on-site use of 360-degree cameras and LiDAR models developed during the "Smart Villages of Italy" practicum fieldwork at MIT (Fall, 2022). By digitally re-creating real places and preserving local knowledge, their platform serves as a model for supporting economic sustainability and creating memory archives of local intangible heritage assets.

Centuripe—a star-shaped hilltop town in the hills surrounding Mt. Etna in Sicily—serves as the investigation site for Karmarkar's participatory mapping effort on vacant buildings (Fig. 9). Karmarkar's research builds upon her prior work in Italy's Abruzzo region, conducted in collaboration with the NGO Liminal, which is dedicated to envisioning actionable, prosperous futures for small settlements. In Abruzzo (Summer, 2022), Karmarkar co-developed a digital data collection tool designed to detect physical signs of abandonment in small-town streetscapes, such as dead plants, major cracks on facades, and large quantities of uncollected mail. In Centuripe, Karmarkar adapted the tool to make it more community-friendly by organizing a dedicated workshop with residents. Participants used the app to visualize and map the extent of abandonment in their town. Beyond the participatory aspect of this project, Karmarkar delves into the spatial statistics of abandonment, revealing clusters of vacant buildings and providing a detailed spectrum of decay. Her findings offer critical insights for local government authorities, helping them address safety concerns in public spaces adjacent to potentially collapsing structures. At the same time, her work provides regional and national governments with a data-driven basis for resource allocation, tailored to the actual conditions of the built environment.

The final contribution of this first section shifts focus from Sicily to Northern Italy, where Fossa, Günther, and Rollandi explore the spatial nuances of the 'digital upgrading' of the Alpine territory. The authors delve into the concept of the 'trans-alpine corridor' by enhancing the linear, valley-shaped topography of Valtellina

Fig. 9 View of the smoking Mount Etna from Centuripe, a star-shaped hilltop town in Enna Province, Sicily. Source: photo by Halim (2022)

which provides the perfect setting for both linear sport activities (such as hiking, cycling, running, ski touring etc.) and cultural routes with monuments and site experience opportunities. Combining new digital accessibility technologies and mobility heritage (e.g., the Bernina railway route and the Via dei Terrazzamenti), the authors argue that visitors in Valtellina can forge a renewed connection with the surrounding landscape, aligned with the principles of slow tourism and inspired by previously developed routes in Switzerland. They propose a Sport&Heritage corridor: an innovative integration between sport activities and heritage fruition, supported by a scenic train line and made possible by dedicated Apps. The Sport and Heritage corridor can become a lasting legacy of the upcoming Olympics, reshaping the future of this international region straddling Italy and Switzerland. Unlike other chapters, Fossa, Günther, and Rollandi adopt a more design-oriented and speculative lens, drawing on historical data and offering reflections informed by contemporary planning efforts across the Alps.

Part II: Remote Working Strategies as Regional Development Frameworks

The COVID-19 pandemic has significantly reshaped the geographies of work worldwide, leaving behind lingering impacts and still-uncertain long-term implications. In this section of *Small-Town Renaissance*, we have grouped contributions that explore how pandemic-induced shifts in how people live, use, and engage with rural areas have evolved in the post-pandemic context. This section focuses on defining a new form of temporary—or even long-term—living conditions in shrinking areas. As the traditionally urban, job-centric spatial nexus described by scholars like Richard Florida comes under scrutiny, this section invites readers to consider whether digitally performed jobs can be effectively conducted from the countryside. It raises critical questions about whether remote workers can integrate into and actively contribute to local communities rather than functioning solely as "extractors" of natural beauty and low living costs as digital nomads.

In the first chapter of section two, Militello—a legal scholar and President of the NGO "South Working"—develops a comprehensive comparative framework examining Southern European countries that have implemented targeted programs to attract remote workers to their most economically disadvantaged regions. Drawing on her on-the-ground experience and a rich dataset profiling "Southworkers," Militello explores the opportunities and risks associated with remote-work-based development strategies. This chapter adopts an interdisciplinary approach to remote work in marginalized areas of Italy, integrating analysis of the scale of the phenomenon, legal challenges related to existing and proposed regulations, and the policy implications of integrating newcomers into small-town communities. Specifically, Militello analyzes a robust dataset comprising 2594 survey responses collected by the Italian nonprofit association "South Working," a map of 230 coworking spaces located in underprivileged regions, and 18 open-ended interviews with individuals who have participated in "South Working" initiatives. Through this rigorous and forward-looking research, Militello offers a series of policy proposals addressing key areas such as connectivity rights, transportation infrastructure, and social amenities, with a particular emphasis on expanding coworking spaces to support sustainable remote work environments (Fig. 10).

Building on Militello's experience and theoretical framework, Agarwal and Wang operationalize a pioneering example of remote-working regional planning for the Valtellina area (Lombardy). By combining interviews with local stakeholders, digitally conducted community workshops, and physical planning, the authors outline a series of actionable regional planning principles aimed at transforming Valtellina into a prime destination for remote workers, fostering a new knowledge economy, and further enhancing sustainable tourism in the region. Agarwal and Wang's work includes a compelling profiling effort, shedding light on the diverse spectrum of stakeholders in rural areas: the tourist, the hybrid worker, the digital nomad, and the local resident. By improving accessibility to Milan, the authors suggest that Valtellina holds exceptional promise as a remote-work hub, offering Milano-based firms the opportunity to explore partial relocation or open secondary

Fig. 10 SouthWorking Castelbuono, a remote-working hub located in a former monastery provided by the municipality of Castelbuono (Palermo, Sicily), serves as the local chapter of the national movement and NGO SouthWorking. This initiative aims to support remote work from areas impacted by brain drain and depopulation. The hilltop town, perched in the Madonie Mountains, exemplifies this effort. Pictured here are Elena Militello (SouthWorking), Brent D. Ryan (MIT), and Fabrizio Barreca (SouthWorking Castelbuono). Source: photo by Ignaccolo (2022)

offices in the Sondrio area. The chapter concludes with design-oriented reflections on the physical spaces conducive to remote work, detailing the types of built environments and outdoor settings that can foster suitable working conditions. It also outlines the range of actors—ranging from national governments to local townships—who can be involved in implementing these plans.

The third and final chapter of this section and the concluding chapter of the book continue the scholarly exploration of the Valtellina area, offering a nuanced contribution at the intersection of slow tourism and digital upgrading. In this chapter, Fossa, Günther, and Paolini introduce the concept of the "Smart Terrace Trail", an innovative approach to enhancing the iconic terraced vineyards of Valtellina. The authors describe how coworking spaces, local gastronomy, and heritage preservation interventions can all be integrated into a unified sustainable and attractive framework for rural landscape revitalization. The focus is on disclosing the potential of ancient paths to support a new form of active healthy mobility integrated with creative working. In their terms, "smart walking" means walking while developing ideas, being inspired by the landscape and supported by available digital connection and plug-in outlets diffused along trails. The smart trail idea offers a physical spine for smart workers' facilities, combining outdoor and indoor spaces, benefiting both residents and tourists alike. This pilot project is aimed at the regeneration of mid-mountain regions (mid-valley regions, often bypassed or abandoned, located below the altitude necessary to guarantee enough snow for ski activities). The low-mid Valtellina can become a model for other regions looking to combine experiential tourism, health mobility, and heritage preservation to support sustainable development along unique natural heritage landscapes.

Technology, Place, and the Future of Small Towns

Amid rising aging populations and ongoing depopulation trends, the scholarly debate on the future of non-metropolitan areas – and their roles in heritage preservation, local production, and land stewardship – remains pivotal. The chapters of this book, along with its broader framing, acknowledge these widely discussed phenomena while taking a nuanced stance by reflecting on them through a technology-focused lens.

In essence, the contributions collected in *Small-Town Renaissance* offer a multifaceted exploration of how digital tools and technologies reshape the socio-cultural and economic landscapes of shrinking areas. These chapters reveal a dynamic interplay between technology, place, and community, challenging traditional notions of scale and presenting innovative pathways for sustainable development. By centering on digital approaches, this volume underscores the potential of new methodologies to address persistent issues in small-town settings while raising critical questions about the future of remote work, heritage preservation, and inclusive planning. However, this book does not delve into the essential yet complex conversation around the ownership and long-term sustainability of these tools. A deeper critical reflection is required to examine the potential downsides or unintended effects of technological ubiquity. For instance, future research on the distribution of remote work facilities or Fixed Wireless Accessibility networks should interrogate what communities these tools benefit and what socio-economic vulnerabilities might be exacerbated by a technology-driven transformation of these landscapes. We hope that the unique empirical and grounded investigations presented here will inspire future research to build on these insights, offering actionable solutions while identifying areas for improvement.

Finally, we invite readers to engage with the localized strategies discussed in each chapter, reflecting on their broader implications for similar contexts across Europe and beyond. Ultimately, *Small-Town Renaissance* seeks to contribute to the growing discourse on rural and small-town revitalization, offering practical lessons and theoretical reflections that can inspire further research and policy innovation in this crucial area of study and strategic implementation.

Acknowledgements The authors thank Fondazione Rocca, MIT Italy, and Fondazione Politecnico di Milano for their support in making this research project and book possible. We are also deeply grateful to those who shared their geographic expertise for this introduction chapter, providing valuable insights into recent trends beyond Italy, including Europe (Deborah Paolini), Switzerland (Annalisa Rollandi), Japan (S. Aika), and China (Quying Wang).

References

Atkinson, C. L., & Atkinson, A. M. (2023). Impacts of climate change on rural communities: Vulnerability and adaptation in the global south. *Encyclopedia, 3*(2), 721–729. https://doi.org/10.3390/encyclopedia3020052

Banda Ultralarga. (2024). *Dati copertura fibra*. Retrieved December 7, 2024, from https://bandaultralarga.italia.it/mappa/

Barca, F., & Lucatelli, S. (2014). *Strategia Nazionale per le aree interne: Definizione, obiettivi, strumenti e governance*. Material UVAL.

Birindelli, A. M., & Balletta, F. (1976). *Un secolo di emigrazione italiana (1876-1976)*. Centro Studi Emigrazione.

Bondolfi, S., Pisani, C., & Rihs, D. (2019). *Vivere e lavorare in montagna grazie a internet*. swissinfo.ch. Retrieved July 26, 2024, from https://stories.swissinfo.ch/vivere-e-lavorare-in-montagna-grazie-a-internet

Bürgin, R. (2021). *Digitalization in the mountains—Digital transformations and digital multi-locality in Swiss mountain areas [Bern]*. Retrieved from https://www.e-helvetica.nb.admin.ch/api/download/urn%3Anbn%3Ach%3Abel-bes-6501%3A21buergin_r.pdf/21buergin_r.pdf

Cabinet Office. (2024). *White paper on aging society 2024 (entire version)*. Retrieved from https://www8.cao.go.jp/kourei/whitepaper/w-2024/zenbun/06pdf_index.html

Cartes de la Suisse—Confédération Suisse. (2024). *Map.geo.admin.ch*. Retrieved from https://map.geo.admin.ch/#/map?lang=de&center=2643525,1200824.8&z=1&bgLayer=ch.swisstopo.pixelkarte-farbe&topic=ech&layers=KML%7C; https://www.dropbox.com/s/dc94x3x8ilf086m/EarthPointExcel_144136.kml?dl=1

Chen, X., Lang, W., Yuan, Y., Yan, G., & Hou, X. (2024). Conceptualizing the nexus between spatiotemporal shrinkage patterns of natural cities and driving mechanisms: Insights into urban shrinkage in Northeast China. *Cities, 152*, 105179. https://doi.org/10.1016/j.cities.2024.105179

Consiglio Federale. (2023). *Comunicato stampa—Il Consiglio federale vuole un accesso Internet veloce per tutta la Svizzera*. Retrieved from https://www.admin.ch/gov/it/pagina-iniziale/documentazione/comunicati-stampa/comunicati-stampa-consiglio-federale.msg-id-99274.html

Cowo. (2024). *Coworking exfabrica Val di Blenio—Torre Blenio—Ticino*. Rete Cowo Coworking Network. Retrieved from https://cowo.it/coworking-network/coworking-ticino-blenio-fabbrica-del-cioccolato/

De Carlo, G. (2004). Speaking of places. *Places Journal, 16*(2), 6.

Deng, T., Wang, D., Yang, Y., & Yang, H. (2019). Shrinking cities in growing China: Did high speed rail further aggravate urban shrinkage? *Cities, 86*, 210–219. https://doi.org/10.1016/j.cities.2018.09.017

Digital Agency. (2024). *Priority plan for the realization of a digital society*. Retrieved July 25, 2024, from https://www.digital.go.jp/policies/priority-policy-program

Eberstadt, N. (2024). *The age of depopulation. Foreign affairs*. Retrieved from https://www.foreignaffairs.com/world/age-depopulation-surviving-world-gone-gray-nicholas-eberstadt

European Commission. (2021). *A long-term vision for the EU's rural areas—Towards stronger, connected, resilient and prosperous rural areas by 2040 [COM(2021) 345 final]*. European Commission.

European Commission. (2024a). *Press release: Report highlights good progress made on long-term vision for EU rural areas*. Retrieved from https://ec.europa.eu/commission/presscorner/detail/en/ip_24_1727

European Commission. (2024b). *Press release—Easier access to EU funding for rural areas thanks to EU rural toolkit*. Retrieved from https://ec.europa.eu/commission/presscorner/detail/en/ip_24_656

European Commission. (2024c). *The long-term vision for the EU's rural areas: Key achievements and ways forward [COM(2024) 450 final]*. European Commission.

European Vision. (2024). *Flagship: Revitalising rural areas most affected by population loss. Rural vision*. Retrieved from https://rural-vision.europa.eu/action-plan/stronger_en#flagship-creating-a-stronger-innovation-ecosystem-for-rural-areas

Eurostat. (2023). *Predominantly rural regions experience depopulation*. Retrieved from https://ec.europa.eu/eurostat/web/products-eurostat-news/w/ddn-20230117-2

Florida, R. L. (2008). *Cities and the creative class (transferred to digital printing)*. Routledge.

Golini, A., Basso, S., & Reynaud, C. (2003). Population aging in Italy: A challenge for our country, a workshop for the world.

He, S. Y., Lee, J., Zhou, T., & Wu, D. (2017). Shrinking cities and resource-based economy: The economic restructuring in China's mining cities. *Cities, 60*, 75–83. https://doi.org/10.1016/j.cities.2016.07.009

Hu, Y., Wang, Z., & Deng, T. (2021). Expansion in the shrinking cities: Does place-based policy help to curb urban shrinkage in China? *Cities, 113*, 103188. https://doi.org/10.1016/j.cities.2021.103188

ISTAT. (2023). *Demographic indicators 2023 (statistiche report)*. ISTAT. Retrieved from https://www.istat.it/wp-content/uploads/2024/03/Indicatori_demografici_26_03_2024_english.pdf

Killé Elsig, C. (2023). Les Haudères: Alpiness coliving accueille des télétravailleurs du monde entier. Visite guidée. *Les Haudères: Alpiness coliving accueille des télétravailleurs du monde entier*. Visite guidée. Retrieved from https://www.lenouvelliste.ch/valais/valais-central/herens-district/evolene-commune/les-hauderes-alpiness-coliving-accueille-des-teletravailleurs-du-monde-entier-visite-guidee-1336198

Kyoto Satoyama SDGs Lab. (2021). *Kyoto Satoyama SDGs Lab*. Retrieved from https://www.kotos-kyoto.jp/646

Li, C. (2004). China's northeast: From largest rust belt to fourth economic engine? *China Leadership Monitor, 9*(9), 1–15.

Li, H., & Mykhnenko, V. (2018). Urban shrinkage with Chinese characteristics. *The Geographical Journal, 184*(4), 398–412. https://doi.org/10.1111/geoj.12266

Li, Z., Wang, X., Feng, Z., Zheng, Y., Wang, J., & Wu, K. (2024). Progress toward some of sustainable development goals in China's population-shrinking cities. *Journal of Cleaner Production, 448*, 141672. https://doi.org/10.1016/j.jclepro.2024.141672

Liu, X., Wang, M., Qiang, W., Wu, K., & Wang, X. (2020). Urban form, shrinking cities, and residential carbon emissions: Evidence from Chinese city-regions. *Applied Energy, 261*, 114409. https://doi.org/10.1016/j.apenergy.2019.114409

Long, Y., & Gao, S. (Eds.). (2019a). *Shrinking cities in China: The other facet of urbanization*. Springer. https://doi.org/10.1007/978-981-13-2646-2

Long, Y., & Gao, S. (2019b). *Shrinking cities in China: The overall profile and paradox in planning* (pp. 3–21). Springer. https://doi.org/10.1007/978-981-13-2646-2_1

Meng, X., & Long, Y. (2022). Shrinking cities in China: Evidence from the latest two population censuses 2010–2020. *Environment and Planning A: Economy and Space, 54*(3), 449–453. https://doi.org/10.1177/0308518X221076499

Ministry of Internal Affairs and Communications, Regional Communications Promotion Division, Information Distribution Administration Bureau, Regional Informatization Planning Office, Local Administration Bureau. (2022). *Reference cases of digitalization of local communities [version 2.0]*. Retrieved July 25, 2024, from, https://www.soumu.go.jp/main_content/000835175.pdf

Ministry of Internal Affairs and Communications, Regional Communications Promotion Division, Information Distribution Administration Bureau, Regional Informatization Planning Office, Local Administration Bureau. (2024). *Reference cases on digitalization of local communities [version 3.0]*. Retrieved July 25, 2024, from https://www.soumu.go.jp/main_content/000949893.pdf

Ministry of Land, Infrastructure, Transport and Tourism. (2022). *Percentage of teleworkers increased from the previous year! The results of the survey on the teleworking population in fiscal 2021 will be released*. Ministry of Land, Infrastructure, Transport and Tourism. Retrieved from https://www.mlit.go.jp/report/press/toshi03_hh_000085.html

Mitsubishi Research Institute, Inc. (2024). *Region Ring®, a digital local currency service that solves local issues*. Retrieved July 25, 2024, from https://www.mri.co.jp/service/platform-services-for-regional-digital-currency.html

Morettini, G., & Chiapparino, F. (2018). Rural 'Italies' and the Great Crisis. Provincial clusters in Italian agriculture between the two world wars. *Journal of Modern Italian Studies, 23*(5), 640–677. https://doi.org/10.1080/1354571X.2018.1535942

Nishi-Izu Town Tourist Association. (2024). Nishiizu Town Sightseeing Guide. Retrieved July 25, 2024, from https://www.nishiizu-kankou.com/

Open Fiber. (2021). *Sustainability report. Open fiber*. Retrieved from https://openfiber.it/app/uploads/2023/06/Sustainability-Report-2021.pdf.

Osaka Pref. (2024). *Subsidy for ICT introduction support project in Osaka Prefecture*. Retrieved July 25, 2024, from https://www.pref.osaka.lg.jp/o090100/koreishisetsu/ict3/index.html

Pezzi, M. G., & Urso, G. (2016). Peripheral areas: Conceptualizations and policies. Introduction and editorial note. *Italian Journal of Planning Practice, 6*(1), 1–19.

Pires de Almeida, M. A. (2017). Territorial inequalities: Depopulation and local development policies in the Portuguese rural world. *Ager - Revista Científica Sobre Desarrollo Rural, 22*, 61–87. https://doi.org/10.4422/ager.2016.08

Police, M. (2023). *Aux Haudères, un nouveau repère pour nomades digitaux*. Canal9. Retrieved from https://canal9.ch/fr/aux-hauderes-un-nouveau-repere-pour-nomades-digitaux/

Ryan, B. D. (2020). Shrinking cities, shrinking world. In T. Banerjee & A. Loukaitou-Sideris (Eds.), *The new companion to urban design*. Routledge.

Sassen, S. (1991). *The global city*. Princeton University Press.

SECO. (2024a). *Politica per le aree rurali e le regioni montane*. Retrieved from https://www.seco.admin.ch/seco/it/home/Standortfoerderung/Regional_Raumordnungspolitik/kore/p-lrb.html

SECO. (2024b). *Sviluppo territoriale coerente*. Retrieved from https://www.seco.admin.ch/seco/it/home/Standortfoerderung/Regional_Raumordnungspolitik/kore.html

UFAG. (2024). *Progetti di sviluppo regionale (PSR)*. Retrieved from https://www.blw.admin.ch/blw/it/home/instrumente/laendliche-entwicklung-und-strukturverbesserungen/laendliche-entwicklung.html

UFAM. (2024). *Spazio rurale*. Retrieved from https://www.bafu.admin.ch/bafu/it/home/themen/thema-landschaft/landschaft-fachinformationen/landschaft-in-der-regionalentwicklung/laendlicher-raum.html

Ufficio Federale dello Sviluppo Territoriale ARE. (2024). *Progetto territoriale Svizzera*. Retrieved from https://www.are.admin.ch/are/it/home/raumentwicklung-und-raumplanung/strategie-und-planung/raumkonzept-schweiz.html

United Nations Department of Economic and Social Affairs, Population Division. (2024). *World population prospects 2024 ((UN DESA/POP/2024/TR/NO. 9))*. United Nations.

UST. (2024). *Attualizzazione 2020 dei livelli geografici non istituzionali - Spazio a carattere urbano, agglomerati e tipologie: Rapporto esplicativo Pubblicazione*. Ufficio Federale di Statistica. Retrieved from https://www.bfs.admin.ch/asset/it/30665780.

Yang, Q., & Ley, D. (2019). Residential relocation and the remaking of socialist workers through state-facilitated urban redevelopment in Chengdu, China. *Urban Studies, 56*(12), 2480–2498. https://doi.org/10.1177/0042098018790724

Yang, Y., Wu, J., Wang, Y., Huang, Q., & He, C. (2021). Quantifying spatiotemporal patterns of shrinking cities in urbanizing China: A novel approach based on time-series nighttime light data. *Cities, 118*, 103346. https://doi.org/10.1016/j.cities.2021.103346

Zhou, K., Yan, Y., & Qian, F. (2019). A multi-scaled analysis of the shrinking population in a region with out-migration: A case study of Hunan province. In Y. Long & S. Gao (Eds.), *Shrinking cities in China: The other facet of urbanization* (pp. 25–41). Springer. https://doi.org/10.1007/978-981-13-2646-2_2

Heritage Management and Enhancement through Digital Tools

Heritage and Mobile Radio: A Data-Driven Approach to Address Cultural Heritage Valorization

Cristina Boniotti and Andrea Pimpinella

Introduction

If transformation is to be an inevitable condition, planning was defined as "the management of change" (Lynch, 1972, p. 1), whereas conservation took shape as "the search for a regulation of transformation that, in the awareness of the uniqueness of each testimony and its multiple documentary character, maximizes permanence, adds its own sign, and reinterprets without destroying" (Bellini, 1996, p. 2). Cities are the subject of continuous change (Ryan, 2017, p. 112), and cultural heritage should not be considered as being static, but rather as a critical mass of evolutionary potential (Della Torre, 1999, p. 75). Thus, on the one hand plural urbanism is primarily intended at understanding the multiple dimensions of places (Ryan, 2017, p. 112); on the other, conservation is not intended at maintaining a building's status quo, but rather managing a mid-to-long-term cultural heritage transition from the static concept of restoration as a one-time event towards conservation as an ongoing process of care (Della Torre, 1999, pp. 74–78). In the light of these propositions, the monitoring and managing of change may also warrant the analysis of heterogeneous and multiple-domain data (i.e., Big Data analytics) providing evidence of people's habits over time and within a specific environment.

The existence of a close relationship between a given anthropogenic environment and the habits of the clusters of humans occurring therein is a widely accepted

C. Boniotti (✉)
Department of Architecture, Built Environment and Construction Engineering, Politecnico di Milano, Milan, Italy
e-mail: cristina.boniotti@polimi.it

A. Pimpinella
Department of Management, Information and Production Engineering, University of Bergamo, Bergamo, Italy
e-mail: andrea.pimpinella@unibg.it

G. Fossa et al. (eds.), *Small-Town Renaissance*, Research for Development,
https://doi.org/10.1007/978-3-031-89786-3_2

matter of fact both from an infrastructural (e.g., transportation systems, leisure and healthcare facilities, digital access, etc.) and land-use (whether residential, commercial, or industrial) standpoint. In the light of mobile radio access ubiquitousness and the widespread use of mobile devices, similar considerations also apply to the communication habits of cellular-phone users—thus, assessing their network activity may shed light on a slew of large-scale social, topological, and technological phenomena (Ayan et al., 2021; Naboulsi et al., 2015; Singh et al., 2019). In fact, data collected either at a user equipment (i.e., via crowd-sourcing) or at a mobile radio access-level (i.e., from counters installed at the radio access points), or even within the core network (e.g., using network probes) is widely employed to monitor large groups of users at a dismal cost not only by mobile operators and for networking and planning purposes, but also by various stakeholders and for scopes spanning several disciplines such as physics, sociology, epidemiology, and urban planning and design. Besides, many studies (Ayan et al., 2021; Lanza et al., 2022; Legeby et al., 2023; Work From Home, 2024) have recently employed data pertaining to user digital activity (e.g., multimedia content downloads and/or social network interactions at given points in time and space) to assess the impact of the COVID-19 pandemic on people's habits, with special focus on transit patterns either on a city-wide (e.g., places of work and/or leisure) or regional (e.g., mobility flows towards places of secondary residence) level.

Data-driven research and big-data analytics are gradually becoming the subject of adoption even within the cultural heritage domain, especially so in the realm of information collection and activity promotion (Bonfantini, 2016; Villa, 2017; Wallace et al., 2023). Besides, research scholars have recently stressed the need for evidence-based policy evaluation and the availability of a "new data landscape" or some degree of innovation in preservation policy (Avrami, 2019). Nevertheless, and to the best of our knowledge, no studies have yet sourced mobile-network data to estimate the number of visitors at any specific cultural heritage site. To investigate the relationship linking cultural heritage to its attractiveness by way of mobile radio data analytics is just the scope of this work.

As far as our project is concerned, inland areas boast some significant cultural heritage, which includes tangible assets (historical palaces and buildings, churches, rural constructions, etc.), intangible assets (local traditions, craftsmanship, religious practices, music, dance, etc.), and landscape (terraced vineyards, mountain pastures, etc.). At the same time, we ought to underscore the difficulties in identifying the exact number of users enjoying this heritage in the light of its extension, fragmentation, and the numerous entities in charge of its management, and, also, the lack of overall data, keeping in mind that the mere number of tickets sold will return highly biased figures. Given that Article 6, Paragraph 1 of Legislative Decree No. 42 of January 22, 2004 (Italian Cultural Heritage and Landscape Code) defines valorization as the "[…] exercising of the functions and the regulation of the activities aimed at promoting the knowledge of cultural heritage and at ensuring the best conditions of use and of public fruition of the asset […]", our contribution is aimed at identifying the way best suited to estimate a given level of public fruition by measuring cultural heritage user attendance and density.

Thus, the main research questions are:

1. *Can mobile network data be of use in assessing the potential for cultural heritage fruition?*
2. *To what degree has the potential of cultural heritage fruition changed in the Valtellina area after the pandemic period?*

This paper thus fits into the broader category of studies aimed at assessing the relationships linking digitalization on the one hand, and cultural heritage support within inland areas, on the other. It is a first-time attempt at investigating the correlation between pre- and post-pandemic cellular phone-user density and the fruition and valorization of cultural heritage for the purpose of providing precious guidance in the definition of novel management strategies.

Our Study will start by elucidating the research methodology adopted and the role of cultural heritage and telecommunications within inland areas during the post-pandemic period. The reasons why we opted for the Valtellina area to begin with will be then exposed by way of reference to its cultural context and the actual funding opportunities on hand. Hereinafter, we will proceed to set forth our mobile data analysis, the dataset we resorted to, the analytical method we followed, our results, their discussion, and a list of research limitations. The final chapter will include some final considerations and reflection as to the instances of future research warranted.

Methodology

The general research methodology we adopted was both of a qualitative and quantitative nature.

The first part of our work was conducted through a theoretical study, by investigating literature related to the role of cultural heritage and telecommunications in inland areas during the post-pandemic period, and the background of the case study selected for the mobile data analysis to be developed subsequently.

The second part was conducted though mobile network data processing in order to assess the potential for cultural heritage fruition. To such effect, a set of data was extracted from the information collected from one of Europe's foremost mobile traffic providers, as well as statistics elaborated on the basis of open-source databases managed by the Italian National Institute of Statistics (ISTAT) and a Data Portal of the Region of Lombardy. Upon pre-processing, our dataset was analyzed to model the potential posed by visiting a Valtellina heritage site through the creation of a "cultural heritage fruition potential index". Finally, heatmaps were elaborated to graphically elicit index-value distribution. In addition, a comparison was drawn between pre- and post-pandemic distribution patterns.

The Post-pandemic Context in Inland Areas

The Role of Cultural Heritage

Under the name of Next Generation Europe, the European Union recently defined a shared strategy aimed at coping with the challenges posed by the COVID-19 pandemic by way of fostering regeneration, sustainability, and innovation (European Commission, 2020a, 2020b; European Commission, Directorate-General for Budget, 2020; European Commission, Directorate-General for Budget, 2021). The twin green and digital transitions are currently the subject of intense promotion (Muench et al., 2022). However, not all European regions are equally well-equipped to master it in terms of digital readiness and growth potential, especially so within rural and agricultural areas (Maucorps et al., 2022).

As part of Next Generation Europe, the project New European Bauhaus has been launched with the aim of establishing a creative and transdisciplinary movement promoting a mutual bond between the realms of science, technology, and culture, and honoring the role of culture in the implementation of sustainable development goals (European Commission, Joint Research Centre, 2022).

Since cultural heritage has been heavily affected by the pandemic, heritage-related organizations have been assessing most of the implications of the COVID-19 crisis within the various dimensions comprising the cultural heritage sector (Europa Nostra, 2020: p. 22; KEA European Affairs, 2020), and subsequently uttered their determination to contribute to Europe's recovery by setting recommendations in support of the heritage sector (Europa Nostra, 2020, p. 11). Besides, the European Heritage Alliance Manifesto highlighted seven ways by which cultural heritage may act as a catalyst for positive change (European Heritage Alliance Manifesto, 2020), while the European Cultural Heritage Green Paper, for its part, stressed its contribution to all key areas of the European Green Deal by issuing recommendations to be addressed by policy makers and stakeholders alike (Potts, 2021). Among the objectives recently defined are the development of strategic spatial planning, place-based investments, and social inclusion (Territorial Agenda 2030, 2020; UNESCO – United Nations Educational, Scientific and Cultural Organization, 2016).

In addition, international-grade agendas have envisioned the pandemic as posing an opportunity for local territories and communities worth some degree of rediscovery.

In the Italian context, by way of example, the Recovery and Resilience Plan (*Piano Nazionale di Ripresa e Resilienza—PNRR*) provides for funding aimed at culture and tourism. In particular, more than EUR 1 billion is appropriated to the so-called "Villages National Plan", which aims at valorizing as many as 250 hamlets occurring in inland areas. Of the above amount:

- EUR 420 million have been assigned to 21 experimental projects (one for each Italian Region), focused on the cultural, social, and economic regeneration of deserted villages (the so-called A-line);

- Additional EUR 580 million have been allocated to the development of "local projects of cultural and social regeneration" aimed at protecting cultural heritage and supporting social and economic sectors (the so-called B-line) within a minimum of 229 individual or aggregated municipalities the population of which does not exceed 5000. The latter strategic line is intended at fostering local attractiveness, repopulation, jobs, public-private partnerships, community involvement, renewable and alternative energy use and conservation, biodiversity and ecosystem protection, and some outspoken care for the issues of climate change, water use, pollution, environmental risks, etc. (Ministero della Cultura, 2021a, 2021b).

In addition, EUR 600 million have been allotted to the conservation and valorization of historical rural architecture and landscapes (Ministero della Cultura, 2021c, p. 4).

The Italian Ministry of Culture is thus recognizing and honoring local cultural and landscape as powerful tools in the pursuance of economic growth (Vitale, 2024).

The Role of Telecommunications

Communication networks are experiencing a worldwide revolution. As reported by Ookla (2024), a company providing connection speed-testing tools worldwide, today's fixed broadband speeds[1] have almost doubled since 2018, moving from an average rate of 45.9 Mbps all the way to 84.3 Mbps. Besides, the most competitive mobile radio access infrastructures are evolving towards 5G and beyond its ecosystem, which today offers median rates higher than 200 Mbps to more than 1.7 billion subscribers (Ericcson, 2024; Ookla, 2023). Considering the crucial role of broadband technologies as carriers of IP traffic, the growth of broadband speed (in both fixed and mobile environments) has been leading to increased consumption and use of high-bandwidth contents and applications. In particular, demand is mostly fueled by the ever-wider availability of multimedia services: for the sake of example, on-demand video streaming is nowadays accountable for almost 70% of Internet traffic,[2] with effects on traffic growth that appear to be exponential rather than linear (Sandvine, 2023). Also, high-bandwidth speeds are essential to supporting other promising applications such as live web-conferencing and cloud storage, thus making the download of large multimedia files as fast as wire transfers from a hard drive. Thus, the enhancement of fixed Internet accessibility, the enhanced reach of

[1] By definition, broadband Internet access refers to data transmission speeds greater than 2 Mbps. Note that broadband networks include the subcategory of ultra-broadband infrastructures, characterized by transmission speeds greater than 30 Mbps.

[2] Network speed vs time to download a video content in high definition (HD): 20 min at 10 Mbps, 9 min at 25 Mbps, only 2 min at 100 Mbps.

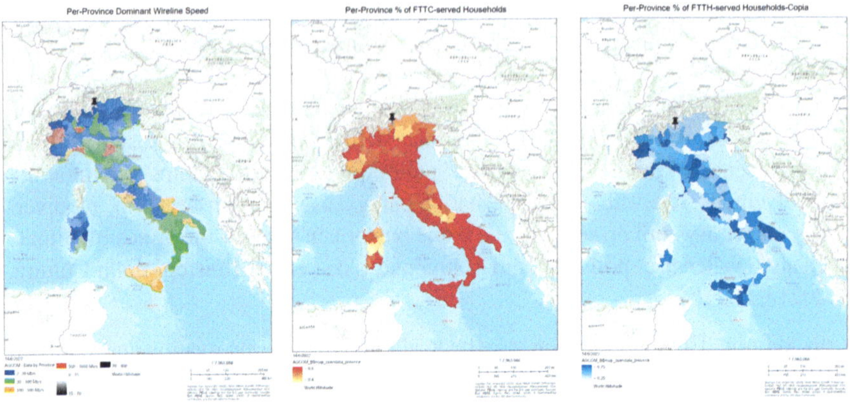

Fig. 1 Color-coded maps of the dominant wireline speed (left), FTTC (center) and FTTH (right) penetration rates, in percentage terms (%) of households. Darker colors are associated with higher penetration rates. Color code for left-side map: blue—[2:30] Mbps; green—[30:100] Mbps; orange—[100:500] Mbps; red—[500:1000] Mbps. Source: image by the authors; data: AGCOM Broadband Map

mobile connectivity to a growing number of end-users,[3] and the growing popularity of Internet of Things (IoT) and Machine2Machine (M2M) applications[4] have set the stage for new networking paradigms and represent the key drivers of a pervasive digital revolution that is bearing—and shall bear—some dramatic impact on our everyday lives.

To provide a comprehensive picture of the Italian wireline ecosystem,[5] we plotted in Fig. 1 (on the left) the by-province distribution of the dominant connection rates with respect to fixed network infrastructure in 2023. Few provinces in Italy (with Milan, Turin, and Rome being among them) can provide to the bulk of their households with wireline connections capable of speeds above 100 Mbps. On the one hand, and considering southern Italy, most of the households experience maximum connection rates ranging between 30 and 100 Mbps, except for Sicily where average top connection rates exceed 100 Mbps for 50% of the households in each province. On the other hand, most of the households living in northern Italian provinces are equipped with broadband rather than ultra-broadband wireline infrastructures, mainly due to the mountain-like morphological characteristics of the territory. Taking the Province of Sondrio (pinpointed by a black marker in Fig. 1) as

[3] To date, mobile users amount to roughly 8.5 billion and are estimated to increase up to 9.3 billion by 2029 (Ericcson, 2024).

[4] E.g., autonomous vehicles, smart metering, video surveillance, healthcare monitoring, transportation, smart car navigation, etc.

[5] On June 3, 2022 AGCOM delivered the online platform "AGCOM BroadbandMap", an open-source database providing data and info-graphics related to the status of the Italian digital infrastructure, including data speed and network coverage of the mobile and fixed access networks operational within the Country.

representative of inland areas, 59% of local households are provided with maximum wireline speeds ranging between 2 and 30 Mbps. Besides the overall broadband access performance, we ought to delve deeper into the distribution of the two fiber optic technologies most widely employed for domestic Internet access,[6] namely Fiber to the Cabinet[7] (FTTC, Fig. 1, mid map) and Fiber to the Home[8] (FTTH, Fig. 1, right map). On the mid and right sides of Fig. 1 we show the color-coded maps of by-province fractions of households served by FTTC and FTTH, respectively. On the one hand, the penetration rate of the FTTC technology in Italy exceeds 42%, serving an average of 80% of the households occurring within a province. On the other hand, FTTH technology shows a much sparser penetration than FTTC: while metropolitan provinces like Milan and Rome hit FTTC penetration rates higher than 89%, on average only 56% of the households in any given province are provided with optic-fiber connections to their apartments. Considering the Province of Sondrio, while 58% of households enjoy the FTTC technology, only 40% of them have access to a FTTH service.

As far as cellular network infrastructure is concerned, Italy boasts one of the highest mobile subscription penetration rates in Europe. According to recent data (AGCOM, 2024), there are more than 100 million active SIMs in the Country, which amounts to a penetration rate of over 150% (most users own more than one SIM). This is mainly due to the extensive coverage made possible by the 4G network available over 99% of the country with average speeds ranging from 25 Mbps within rural areas all the way to 100 Mbps within urban environments. Moreover, the 5G network is expanding, too. As of 2024, and even though over 70% of the population has access to 5G services, only 20% of Italian mobile users own 5G-compatible devices, with major cities and urban areas being better equipped than rural and less densely populated areas, wherein 5G coverage still lingers. Nevertheless, and although many users still rely on the 4G network—especially so within areas still not covered by 5G network – the number of 5G subscriptions is on the increase.

5G and related technologies such as the Fixed Wireless Access (FWA) are deemed to be among the most highly feasible paths towards digitalization for the purpose of revitalizing inner areas in synergy with fixed infrastructure enhancement. Among the many opportunities unleashed by digitalization, we believe that

[6] The architecture of an optical access network is usually referred to as Fiber-To-The-x (FTTx). It consists of two building blocks, namely (1) the Primary Network connecting principal optical nodes to secondary nodes, and (2) the Secondary Network connecting secondary nodes to end users.

[7] In this case, the secondary nodes are distribution cabinets used to deliver copper-based Internet access to end users. Thus, optical cables come to an end at a distribution cabinet level and at distances ranging between a few hundred meters from end-user apartments within urban scenarios up to 2 km within rural scenarios, furthermore, copper cables are used to deliver Internet traffic to an end-user. Maximum connection speeds: 50–100 Mbps *Down*, 1–10 Mbps *Up*.

[8] In this case, the secondary nodes are passive optical splitters that deliver full optical Internet access all the way to individual apartments. Maximum connection speeds: 1000–10,000 Mbps *Down*, 50–100 Mbps *Up*.

access to next generation communications technologies bears the potential of outlining a cultural heritage valorization scenario which often occurs in a rich, yet highly dispersed pattern outside of the infrastructural thoroughfares of the valley bottoms.

Valtellina as a Case Study: The Reasons

Context of Reference

Valtellina is a mountainous area located in the heart of the Italian Alps and within the Region of Lombardy. It may well be considered a marginal, isolated area, as some of its little towns are quite far from most main cities and the reach of essential services such as health, education, and public transportation. Inland areas have typically undergone gradual abandonment starting after World War II. Nevertheless, these places are often characterized by peculiar environmental resources such as agricultural and water systems, extensive green areas, natural reserves crucial for biodiversity, as well as landscapes and cultural heritage sites (Barca & Lucatelli, 2014; Tantillo, 2019).

Indeed, Valtellina boasts some significant cultural heritage thanks to its history and tradition, with the earliest traces of human presence dating as way back as prehistory, as proven by the rock carvings of Grosio and Teglio. Valtellina has been the main gateway to the Alps for centuries. By virtue of its position, it was an important transit route between the north and south of Europe and part of a thoroughfare allowing for merchants to travel from rich Italian ports to the most important cities of Central Europe. It has played a relevant role since ancient Roman times, with Roman Emperor Augustus decreeing the strategic relevance of the alpine thoroughfares crossing the Spluga and Settimo passes. Numerous ancient populations have set their sights on this area and occupied it over the centuries: Lombards (568 A.D.) and Franks, followed by Charlemagne and the City of Como, which in turn consolidated the above routes, and, finally, the Viscontis and Sforzas of Milan (mid-fourteenth century), the latter domain having subsided to the pressure originating from the great powers of that time, i.e., France, Spain, the Republic of Venice, Germany, and Switzerland. Indeed, and starting from 1512, Valtellina became part of the "Three Leagues" of the Swiss canton of the Grisons, and was ruled by its Protestant government, thus playing a strategic role in-between the Latin-Catholic sphere of influence on the one hand and the Germanic-Protestant on the other. In 1815, the Valley was incorporated in the Lombard-Venetian Kingdom, a part the Austrian Empire in those years. From that point in time onwards, the history of Valtellina became one with that of Italy, the unification of which dates back to 1861 (Benetti & Guidetti, 1990; Mariotti, 2016).

In addition, Valtellina was chosen as a pilot case for the project as it has so far attracted a significant amount of public and private investments in both cultural heritage and telecommunications, as shown in the following section.

Heritage-Related Funding

Among the most recent events, we ought to mention the catastrophic floods that struck Valtellina in July 1987 causing dozens of casualties. In addition, the mountain sides underwent severe erosion which in turn begot significant changes in the local landscape. The population was evacuated from their homes and spent endless days in temporary shelter conditions (Benetti & Guidetti, 1990). In response to this catastrophic event, Law no. 102 of 1990, also known as the "**Valtellina Statute**", was issued to fund the reconstruction of the area. In addition, it appropriated EUR 35 million aimed at the restoration of over 70 historical buildings, 25% of which consisting of local funds. We ought to stress that no relevant economic or social effects could be appreciated as a consequence of this initiative, and that no ongoing, long-term conservation interventions were enacted except for the completion of the actual restoration operations (Foppoli, 2018).

The **Cultural District of Valtellina** is one of the most significant heritage projects recently developed in the area. In 2005, the banking foundation Fondazione Cariplo launched the Lombardy-based *Distretti Culturali* (cultural district) project in order to plan for a long-term cultural heritage valorization strategy based on intervention quality and innovative actions, integration of the cultural supply chain within the local economy, and implementation of an innovative activity management approach to be agreed upon in conjunction with major local stakeholders. After a tender and the ensuing, long evaluation process, six proposals pertaining to six different areas were selected for the development of the Cultural Districts, each of which led to the development of different initiatives (Barbetta et al., 2013; CHCfE Consortium, 2015, p. 197; Della Torre, 2015).

The Cultural District of Valtellina attempted to consolidate the relationships between landscape, traditional production, and cultural identity by valorizing both historical buildings and agricultural-culinary traditions.

The District's ground surface amounts to about 2760 m^2, with its population amounting to approximately 155,000 inhabitants distributed in in a non-homogeneous way.

The initiative was managed by the Local Development Foundation (*Fondazione di Sviluppo Locale*) in its capacity as a project leader and in partnership with additional public entities such as the Province of Sondrio, four mountainous-area communities (Morbegno, Sondrio, Tirano, Alta Valtellina), 65 local Municipalities, a few Swiss local administrative bodies, the Chamber of Commerce of Sondrio, several private associations and operators, some local banks, and an electric power company.

The project was developed between 2010 and 2015 and was granted a contribution of EUR 8,290,000 by virtue of a co-financing scheme involving both the Cariplo Foundation and several local administrative bodies, with the stakes of the above foundation amounting to a rate of 57%.

The objectives encompassed conservation of cultural heritage sites, skill and expertise building, and entrepreneurial innovation promotion, with special care for the food and beverage sectors, all of which while aiming at eliciting some active involvement on part of the local community. Special attention was also devoted to educational activities aimed at local professionals, farmers, craftsmen, teachers, and students which revolved around the topics of cultural heritage, landscape, and the so-called "preventive and planned conservation" approach, and was implemented by way of courses, workshops, as well as events and contests. Quite surprisingly to most stakeholders, attendance to the above courses and conferences exceeded the 3000-participant threshold.

As to the peculiar issue of conservation interventions, some historical buildings were restored, while an assessment was performed on the buildings that had undergone a Valtellina-Statute restoration in the previous years. Indeed, most of the latter buildings did not undergo any proper inspection and long-term maintenance after restoration completion. Different activities including the execution of preliminary diagnostic analyses and the drafting of conservation plans were carried out on the Saint Antonio Cloister in Morbegno, the Castle of Masegra in Sondrio, and the Tirano Railway Station, with the latter of which being part of the UNESCO buffer zone going by the name of "Rhaetian Railway in the Albula/Bernina Landscapes" (Foppoli, 2018; Foppoli et al., 2019; Foppoli & Di Capita, 2018).

Among the plans developed within the Valtellina Cultural District, "The Route of Terraces" (*Via dei Terrazzamenti*) is a path aimed at enhancing the terraced vineyards carved out of the steep and rocky slopes of the Rhaetian Alps with the use of dry-stone walls. The largest artefact of that ilk on Italian soil, they cover over 850 hectares and account for 37% of the overall terraced vineyard surface nationwide. By making possible this initiative, locals managed to foster their economy by transforming the pre-existing natural environment, which in turn happened to result in the need for ongoing care and maintenance. A 74 km-long compound of pre-existing mountain paths and trails, the Route of Terraces is a pedestrian and cycling path connecting Morbegno to Tirano, to wit, the lower and the median sections of the Valley, which highlights most of the cultural sites, cellars, and villages located along its way as well as many a local artisanry shop. Various stakeholders were involved not only in the planning stage of this route, but also the actual management thereof. Besides, an instance of close collaboration was established as part of the promotional initiatives regarding the Swiss-based "Via Storia" historical route society. Therefore, the "The Route of Terraces" was annexed to the pre-existing "Via Valtellina", an historical route through which local wines were exported to Northern Europe by way of the Swiss Canton of Graubünden and the Austrian Region of Vorarlberg. This link over Italy, Switzerland, and Austria also serves as a connection between Lakes Bodensee and Como.

In light of the above considerations, the landscape characterizing the Valley proved to be a catalyst for the establishment and strengthening of partnerships; besides, these experiences have proven collective efforts as being conducive of the recognition, protection, and conservation of any given cultural landscape (Foppoli, 2018; Foppoli et al., 2019; Vitale, 2019).

The outcomes begotten by the aforementioned Valtellina Cultural District led to the development of a slew of additional heritage projects, to wit:

- The "**Planned Conservation in the Rhaetian Common Space—CPRE**" (*Conservazione Programmata dello Spazio Comune Retico—CPRE*) project (2013–2015): bankrolled by the European cross-border Italy-Switzerland Interreg cooperation program and costing EUR 1,570,000, 56% of which self-funded, the CPRE was aimed at coordinating conservation activities on cultural heritage and landscapes across Valtellina, Italy and the Valposciavo, Switzerland, with its activities revolving around the mainstays of knowledge, conservation, training, and valorization (Foppoli, 2018).
- The "**+++ More Positive Signs**" (+++ *Più Segni Positivi*) project (2014): funded by the Fondazione Cariplo banking foundation by means of the "Welfare in Action" call and planned by public entities in conjunction with social cooperatives. In the light of the social and hydro-geological weaknesses typically characterizing mountainous areas, the project aimed at providing additional job opportunities in the fields of path and terracing maintenance, thus providing protection to the cultural landscape of Valtellina (Foppoli, 2018; Più Segni Positivi, 2022).
- The "**Conserving and valorizing the cultural landscape of mid-Valtellina**" (*Conservazione e valorizzazione del paesaggio culturale della Media Valtellina*) project (2018): funded both by the Fondazione Cariplo banking foundation with a contribution of EUR 900,000 and the Region of Lombardy with EUR 900,000 (total investment: EUR 1.8 million), it was a broad-scope project promoted by the Tirano mountainous-area community together with twelve additional municipalities (Foppoli, 2018; Paesaggi della Media Valtellina, 2023).
- The "**Cultural Heritage and Cultural Landscape of Valtellina and Valposchiavo in the UNESCO Context – ConValoRe**" (*Conservare e valorizzare nello spazio comune retico i beni culturali ed il paesaggio culturale di Valtellina e Valposchiavo nel contesto UNESCO*) project (2020–2023): funded by the European cross-border Italy-Switzerland Interreg cooperation program, costing EUR 1,308,832.07, and implemented by the Municipality of Tirano in its capacity as a project leader together with further partners such as public universities and social cooperatives. It is aimed at enhancing the historical, social, and cultural links between the median section of the Valtellina and the Valposchiavo (Interreg Italia-Svizzera, 2024).

Telecommunications-Related Funding

Italian authorities are leading the Country towards a radical evolution of its digital infrastructure, and doing so in spite of the criticalities embedded in both the spatial and socio-economic conditions of Italian urban settlements. Differently from past decades, during which larger metropolitan areas were extensively equipped with advanced digital infrastructures while rural towns were often left behind, unprecedented resources will be allocated in the years to come to fulfil the digital transition of the Country, the urgency of which has become even more apparent in the wake of the COVID-19 pandemic.

A clear example of this trend is represented by the Recovery and Resilience Plan,[9] a EU-funded plan that will devote roughly EUR 60 billion over 3 years to promote both the drastic reduction of the digital divide[10] and the distribution of a common digital knowledge among citizens, such as to include all of them into one digital environment. As a matter of fact, the exponential growth in the number of Internet accesses, due mainly to the popularity of online services as enablers of remote working and learning, has dramatically remarked the need for ultra-fast, resilient fixed and mobile network infrastructure. Despite the overall decreasing trend in revenue from consumer plans[11] following the pandemic (though a slightly positive increase of 0.8% was observed at the end of 2023), the number of fixed broadband accesses throughout Italy has increased to 18.9 million, up 18% from 2019. In particular, the average number of broadband and ultra-broadband wired lines per 100 residents equals 32.1, with 27.5% and 23.6% of residents enjoying connection speeds in excess of 30 Mbps and 100 Mbps, respectively. Interestingly, focus on the growth of ultra-broadband access infrastructure has mostly revolved around areas like the Province of Sondrio: 92% of newly installed access points in 2023 (amounting to over 730,000) occurred in inland areas (AGCOM, 2024), thus testifying to some remarkable effort in infrastructure development and investment for the purpose of a more homogeneous and inclusive coverage nationwide.

Cultural Heritage Fruition Potential: Data Processing and Analysis

This section presents the approach used to process mobile network data and evaluate how promising access to the system of cultural heritage sites is in the region of Valtellina. Firstly, we will provide a description of the dataset used for this section. We will then proceed to provide details about the pre-processing operations applied to the data, only to finally introduce the Cultural Heritage Fruition Potential, an index designed to represent the potential for a Valtellina-based piece of cultural heritage to be visited.

[9] The overall fund amounts to EUR 222 billion, 27% of which has been allotted to the enhancement of the Country's digital infrastructure, as an add-on to the resources already allocated as part of the Ultra-Broadband National Plan (2015).

[10] "The term "digital divide" refers to the gap between individuals, households, businesses and geographic areas at different socio-economic levels with regard both to their opportunities to access information and communication technologies (ICTs) and to their use of the internet for a wide variety of activities" (OECD, 2001, p. 5).

[11] The overall market value of communication services has decreased by 10% compared to 2019, with losses mostly due to the market dynamics of mobile rather than fixed-connection communications.

Dataset and Pre-processing

This work employs two different sets of data. The first has been provided by one of the most popular European mobile operators, the Italy-wide market share of which exceeds 27% and almost uniformly reflects all different age and economic user segments (AGCOM, 2024). The second dataset, for its part, has been retrieved by the authors from open-source databases managed by the Italian National Institute of Statistics (ISTAT[12]) and a Data Portal of the Region of Lombardy.[13] Hereinafter, we will provide details about the information contained in both datasets.

Cellular Operator Database The first dataset has been collected at the access of the LTE[14] cellular network currently operational in Italy over two 3-week periods, namely P_1 = {20/01/2020, 10/02/2020}, i.e., immediately before the pandemic breakout in Italy[15] and P_2 = {24/01/2022, 13/02/2022}, i.e., when most of the COVID-19 related restrictions enacted by the national government had expired. Also, we will restrict our focus to the network section of Lombardy,[16] the fourth largest Italian region by surface area, which comprises Valtellina and hosts over 2500 cellular base stations[17] (known in the LTE jargon as eNodeB's).

The dataset used in this work contains real-world Key Performance Indicators (KPIs) measured independently by each individual eNodeB under scrutiny. Data are collected through ad-hoc counters in the form of hourly sampled time series, and provide information such as cumulative downloaded and uploaded Internet data volumes, the share of busy radio resources, or the number of users connected to a given radio access site. In this work, we will focus on the latter KPI, and report the number of users connected to each eNodeB by way of Radio Resource Control (RRC). In a nutshell, a cellular user can effectively access cellular services only after the establishment of a RRC connection with the network (most likely so with the physically closest eNodeB), that ensures the user is assigned the set of radio signal spectrum needed to send or receive data. Differently from traditional Call Detail Records (CDR), where a record is generated only when a connected device interacts with the network, our KPI counts the number of users who might exchange data with the connected eNodeB every given hour and regardless of whether they

[12] https://www.istat.it.

[13] https://public.tableau.com/app/profile/polis.lombardia/vizzes.

[14] LTE stands for Long Term Evolution and is the technical name used to represent the well-known fourth Generation (4G) technology for mobile radio access.

[15] The first recognized COVID-19 case in Italy dates back to February 9, 2020.

[16] It might happen that a user physically located at the border of Valtellina connects to a cellular access point outside Valtellina. To take this issue into account, we will consider the entirety of the Lombard cellular network.

[17] A base station consists of a transmission antenna and the equipment used for processing radio signals. It is the physical element providing end-users with access to cellular infrastructure, this being the reason why base stations are usually mounted on roof-tops and tilted towards the ground, such as to increase end user coverage.

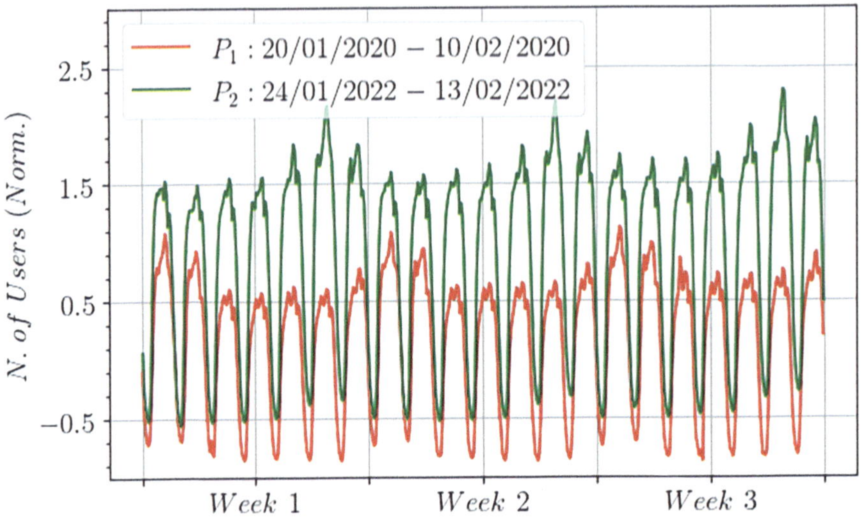

Fig. 2 Normalized hourly number of connected users, during P_1 (red) and P_2 (green). Source: image by the authors; data: MNO

are active (i.e., effectively transmitting) or not, which thus provides a reasonably reliable picture of people presence within the Valley.

Figure 2 plots the aggregate amount of connections reported in Valtellina along P1 (red) and P2 (green), respectively. Out of privacy concerns, all data have been normalized to mean and variance. As already observed in Pimpinella et al., 2024, we ought to note that *the overall presence of mobile users in Valtellina has increased by an average of 30% after the COVID-19 pandemic. Also, we ought to stress two further types of trends. On the one hand, the weekly number of users connected to the network increases as we approach the weekend during both P1 and P2. On the other, the number of connections increases between the first and third week, a tendency more evident during P2 than P1. The latter two are both seasonal effects: Valtellina is in fact a popular winter sport resort.*

Open-Source Databases The second dataset contains descriptive indicators and official statistics about census, administrative management, and tourism intensity relative to all the municipalities in Lombardy, including those located in Valtellina. Precisely, we consider the following information:

- Administrative boundaries, surface area, and urban footprint (i.e., fraction of the administrative area hosting any type of construction, e.g., buildings, streets, etc.) of each municipality located in Lombardy (source: ISTAT).
- Number of residents in each municipality located in Valtellina, as measured in 2020 (source: ISTAT).
- Number of cultural heritage sites per Valtellina-based municipality, as measured in 2017 (source: Data Portal of the Region of Lombardy).

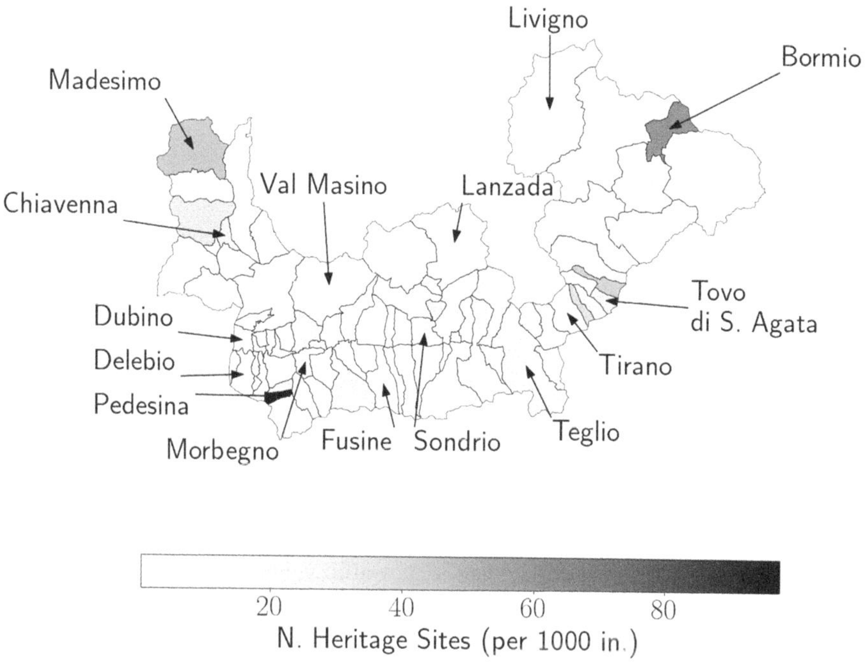

Fig. 3 Map of the Province of Sondrio. The administrative area of each municipality is color-coded according to the number of cultural heritage sites per 1000 inhabitants. Source: image by the authors; data: ISTAT and Data Portal of the Region of Lombardy, elaborated by the authors

- Synthetic index of intensity and type of demand for tourism associated to each municipality located in Valtellina (source: ISTAT). Each municipality is assigned a score from 0 to 5 depending on the number of yearly tourist presences, with 0 identifying non-touristic municipalities and 5 being consistent with instances of extremely high tourist density. The latest updates of this index were delivered in 2022.

In Fig. 3, we plot a grey scale coded map of the number of cultural heritage sites (normalized to 1000 inhabitants) identified within each and every municipality of the Province of Sondrio. *We observe that Pedesina, Bormio, and Madesimo*[18] *have the highest density of cultural heritage sites per 1000 citizens. A slightly lower density is seen in Tirano, a city well-known for hosting the UNESCO World Heritage Site of Bernina Express railway. Among those having the least density values, we*

[18] The Municipality of Madesimo is situated in Valchiavenna, a small valley adjacent to Valtellina that is home to roughly 14% of the population of the Province of Sondrio. Given that both valleys share similar characteristics in terms of orography, infrastructure, and tourism demand, and that Valtellina hosts the remaining 86% of the Province's population, we decided to include the municipalities of Valchiavenna in our analysis.

should mention the municipality of Livigno, a destination often selected by both national and foreign tourists for their winter holidays.

Data Processing

In this section, we will provide details about the approach adopted to elaborate the information retrieved from both datasets, with the final goal of providing some insight as to the potential for a cultural heritage site to be visited by a user physically occurring in Valtellina.

Data Processing Let $u_1(e,t)$ and $u_2(e,t)$ be the number of users connected to the e-th eNodeB at hour t during P_1 and P_2, respectively. Without loss of generality, we can represent the radio coverage area of the e-th eNodeB by its Voronoi polygon V_e. As shown in Fig. 4, a Voronoi mapping is the tessellation of a geographic area that can be computed from a set of source node locations (in this work, the eNodeB's deployed in Lombardy), associating each spatial point to the closest source node. In this way, all cellular network metadata (e.g., traffic volume, number of connected users, etc.) measured by one eNodeB can be uniformly spread over the surface of the corresponding Voronoi cell (Martinez-Durive et al., 2022).

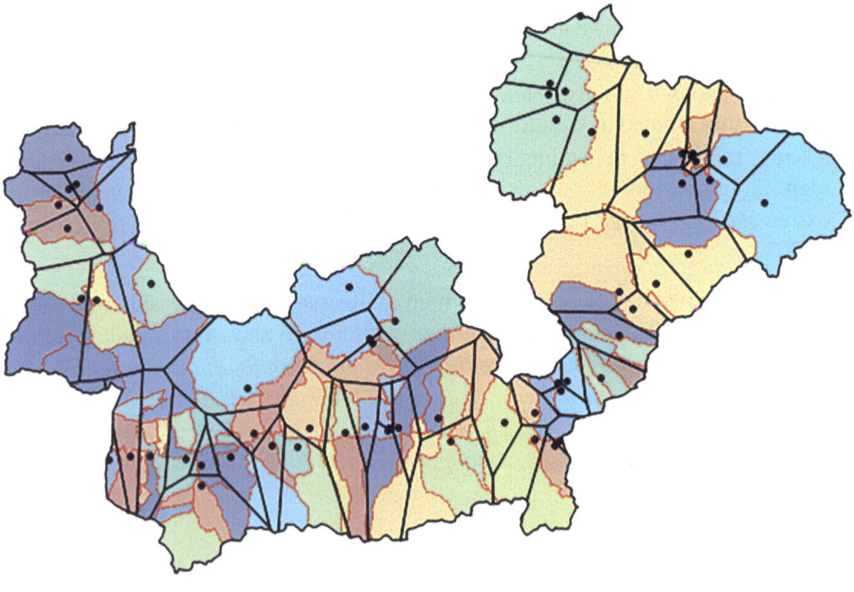

Fig. 4 Map of the Province of Sondrio. Dashed, red lines delimitate the administrative boundaries of the municipalities of the area, while black markers indicate eNodeB locations. Black lines are used to outline the coverage area of each eNodeB as approximated herein by way of Voronoi polygons. Source: image by the authors; data: ISTAT, elaborated by the authors

In light of the scope of this work, we will first process the available data to estimate the number of users connecting from each municipality m of the Valley, during P_1 and P_2 respectively. To do so, we will proceed as follows:

1. First, for each couple (V_e, m) of Voronoi cell and municipality, we compute α_m^e as the area of the intersection between V_e and the administrative area of m, normalized to the area of cell V_e. Thus, α_m^e is a number taking values between 0 (i.e., the administrative area of m has no overlap with cell V_e) and 1 (i.e., the cell V_e is fully contained within the administrative borders of m, or vice versa).
2. Second, we will compute the normalized urban footprint η_i^e that is contained in cell V_e and administratively belongs to municipality m as:

$$\eta_m^e = \cdot \frac{A_m \cdot f_m \cdot \alpha_m^e}{\sum_m \left(A_m \cdot f_m \cdot \alpha_m^e \right)} \tag{1}$$

with A_m being the surface of the administrative area of m (measured in square kilometers) and f_m a number between 0 and 1 accounting for the urban footprint relative to m. So, the numerator $A_m \cdot f_m \cdot \alpha_m^e$ equals the surface in square kilometers of the urban footprint of any m municipality falling within the Voronoi cell associated to the e-th eNodeB.

In other words, all municipalities not covered by eNodeB e (i.e., for which α_m^e is equal to zero) do not contribute to the urban footprint contained in cell V_e. Note that the numerator is then normalized to the sum of the different contributions, i.e., η_m^e takes values between 0 and 1 such that $\sum_m \eta_m^e = 1$ for each considered eNodeB. Also, note that in this calculation we assume that the urban footprint of m is spread over its whole administrative area.

3. Finally, we define the estimated number of users $\widehat{u_1}(m,t)$ and $\widehat{u_2}(m,t)$ that are using the cellular infrastructure at hour t while being physically located in municipality m during $P_1$and P_2, respectively, as:

$$\widehat{u_T}(m,t) = \sum_e u_T(e,t) \cdot \eta_m^e,\ t \in P_T, T = 1,2 \tag{2}$$

Note that the estimated KPI corresponds to the weighted sum of the users connected to all those sites that cover municipality m (i.e., for which $\alpha_m^e > 0$), where the weights η_m^e accounts for the likelihood that a user connected to site e physically occurs in m (i.e., within its urban footprint). In other words, we exclude that a user is connecting from any location within a municipality's unbuilt surface (e.g., forests, rivers, mountains, etc.).

At this point, we leverage the well-known weekly periodicity of cellular network metadata to build a compressed representation of $\widehat{u_1}(m,t)$ and $\widehat{u_2}(m,t)$. A popular approach often used to summarize the usage behavior of an eNodeB is by means of its Median Weekly Signature (Pimpinella et al., 2022). With the same approach, we compute the Median Weekly Signatures (MWS) corresponding to each municipality m, namely $\vartheta_1(m,t)$ and $\vartheta_2(m,t)$. By fixing the reference period P_T, $T = \{1,2\}$, and municipality m, we compute $\vartheta_T(m,t)$ as follows:

1. We calculate the Median Daily Signature (MDS) for each day of the week (i.e., Monday through Sunday). In a nutshell, the MDS for a given day (e.g., Monday) is determined on an hourly basis by obtaining the median value of the estimated number of users connecting from m each hour for every day of the same type (i.e., Mondays) within the period of choice.
2. Next, we determine the MWS by connecting the 7 instances of MDS obtained, one for each day-type. As an example, Fig. 5 depicts the normalized version of $\vartheta(1, \mathrm{Sondrio}, \mathrm{t})$, i.e., the MWS computed for the city of Sondrio with respect to P_1, normalized to mean and variance: as one can see, the MWS remains consistent from Monday to Friday, while a noticeable shift in pattern can be observed over the weekend.

Repetition of this process for each Valtellina municipality, and independently for both P_1 and P_2, allows us to assess the potential of a given cultural heritage site to be visited by a user physically occurring in Valtellina using the MWSs as an input, as described in the following section. Note that such an approach greatly reduces the amount of data to be considered for further analysis, as it lets us work with only 168 data samples (one per hour composing a week), per each considered period and municipality.

Cultural Heritage Fruition Potential Let R_m be the number of residents of municipality m, and C_m be the number of cultural heritage sites located in municipality m. We define the Cultural Heritage Fruition Potential (CHFP) index $\sigma_T^{WD}(m)$ and

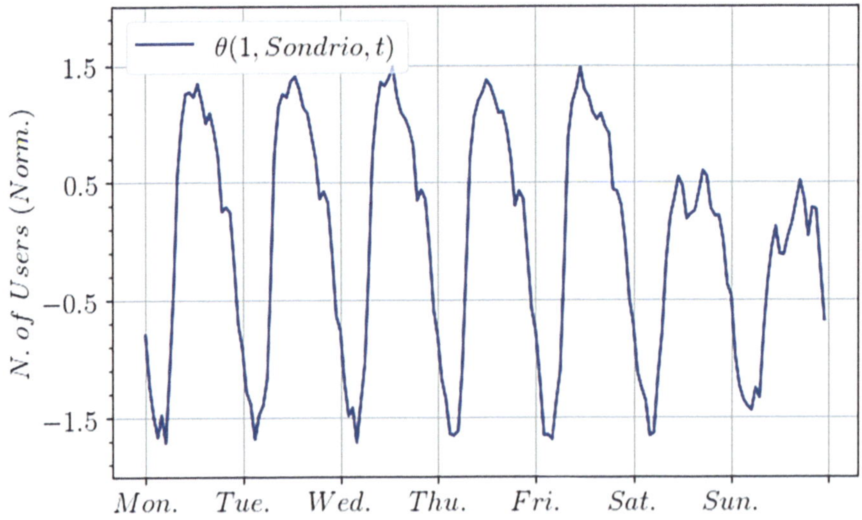

Fig. 5 MWS associated to the municipality of Sondrio during P_1: the MDSs of each day from Monday to Sunday are horizontally stacked to form the MWS. Data have been normalized for privacy reasons. Source: image by the authors; data: MNO, elaborated by the authors

$\sigma_T^{WE}(m)$ associated to municipality m during period P_T, for $T = \{1,2\}$, respectively for WorkDays (WD) and WeekEnd (WE), as:

$$\sigma_T^{WD}(m) = \frac{1000}{R_m} \cdot \frac{C_m}{\sum_m C_m} \cdot \frac{1}{5} \cdot \sum_{\substack{t \in [8\,a.m.,\ 7\,p.m.] \\ \forall day\, d \in \{Mon.,\ \cdots,\ Fri.\}}} \vartheta_T(m,t) \tag{3}$$

$$\sigma_T^{WE}(m) = \frac{1000}{R_m} \cdot \frac{C_m}{\sum_m C_m} \cdot \frac{1}{2} \cdot \sum_{\substack{t \in [8\,a.m.,\ 7\,p.m.] \\ \forall day\, d \in \{Sat.,\ Sun.\}}} \vartheta_T(m,t) \tag{4}$$

Therefore, $\sigma_T^{WD}(m)$ and $\sigma_T^{WE}(m)$ equal the average daily number of estimated potential visitors of a cultural heritage site located in the m-th municipality per 1000 residents, for a work or a week-end day during period P_T, respectively. The two quantities take into account:

1. the frequency of cultural heritage sites located in m with respect to the overall number of cultural heritage sites in Valtellina, i.e., $C(m) / \sum_m C(m)$, and
2. the average number of estimated cellular line users connecting on a single day from municipality m during daily hours alone (i.e., $t \in$ [8 a.m., 7 p.m.) of either workdays (i.e., {Mon., …, Fri.}) or weekends (i.e., {Sat., Sun.}).

In other words, if $\sigma_T^{WD}(m)$ or $\sigma_T^{WE}(m)$ are high, then (1) many users have on average connected during respectively a workday or a weekend day from municipality m and (2) the frequency of cultural heritage sites located in municipality m is high. We underline that $\sigma_T^{WD}(m)$ and $\sigma_T^{WE}(m)$ refer to potential rather than actual visitors as nothing can be inferred from our data about the true reason why a cellular user occurs in municipality *m*\n. In the next section, we will show the results obtained from the calculation of the proposed index, commenting on the differences observed among the different municipalities located in Valtellina.

Results and Discussion

We shall now use the CHFP to assess and compare the potential with which a cultural heritage site is visited by a cellular user connecting from Valtellina, with the goal of answering questions Q1 and Q2. Focusing on periods P_1 and P_2, we will start by leveraging the Synthetic Index of intensity and type of Touristic Demand (SITD) provided by ISTAT to validate $\sigma_T^{WD}(m)$ and $\sigma_T^{WE}(m)$ (Eqs. (3) and (4), respectively) with respect to their ability to reflect the potential of cultural heritage site fruition. After that, we will proceed to use heatmaps to graphically show the distribution of CHFP values among the municipalities located in the Valley, comparing how such distribution has changed between P_1 and P_2. Finally, we will discuss the limitations

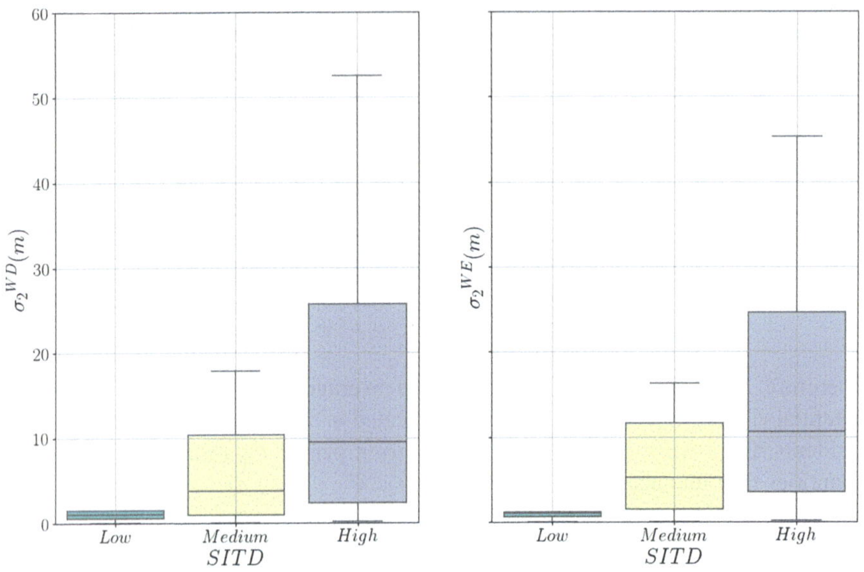

Fig. 6 Distribution of σ_2^{WD} (left) and σ_2^{WE}(right), conditioned to the Synthetic Index of intensity and type of Touristic Demand based class associated to the municipalities of Valtellina. Source: image by the authors; data: MNO, ISTAT, and Data Portal of the Region of Lombardy, elaborated by the authors

to our approach. For the sake of clarity, we ought to stress that all results shown hereinafter are statistically normalized to a population of 1000 residents per municipality.

Validation To validate the extent to which the CHFP is suited to reflect the potential of cultural heritage site fruition, we will consider the Synthetic Index of intensity and type of Touristic Demand (SITD) provided by ISTAT. To this effect, we ought to note that the higher the SITD index of a municipality, the higher the intensity of touristic demand and, thus, its tourist attractiveness. To ease the interpretability of the results, we will further classify each municipality according to the associated SITD value as subject to *Low* (SITD equal to 0), *Medium* (SITD between 1 and 3), or *High* (SITD equal to 4 or 5) tourist occurrence intensity. In Fig. 6, we thus plot the distributions of $\sigma_2^{WD}(m)$ (left plot) and $\sigma_2^{WE}(m)$ (right plot), conditioned to the class assigned to each municipality. The black horizontal line in the boxplots represent the median value of each distribution, while the upper and lower whiskers represent the minimum and maximum values. As one can see, a relation exists between the CHFP and the intensity of tourist demand as to a given city. Both 50-th and 75-th percentiles as well as the maximum values of the two distributions show an increasing trend with respect to such demand. Moreover, while 50% of the municipalities classified as *High* show CHFP values greater than 10 users per day, this is true only for 25% of those classified as *Medium*. Also, none of the latter show CHFP values greater than 19 users per day, while this happens for more than 40%

of *High* municipalities, during both workdays and weekends. Instead, CHFP values lower than 5 users per day are shown by *Low* municipalities, as they either do not host any cultural heritage sites, or both user occurrence frequency and cultural heritage sites are low with respect to other municipalities of the Valley.

In the light of the above observations, we can finally provide an answer to question *Q1*: *it is possible to leverage mobile radio data to provide insight as to the attractiveness of a given municipality and assess the potential for cultural heritage fruition. To assess the attractiveness of the different municipalities located in Valtellina, we performed comprehensive data post-processing and joint analytics of cellular and public entity-related data, thus leading to the design of the Cultural Heritage Fruition Potential index. We validated CHFP showing its consistency with independent public data relative to tourist demand intensity within the Valley. Such results confirm that the habits of cellular users in terms of network activity reflect the characteristics of the underlying urban area in which such activity takes place.*

On the Variation of CHFP Values from 2020 to 2022 We plot in Fig. 7 four heatmaps of the Province of Sondrio, where each municipality is color-coded according

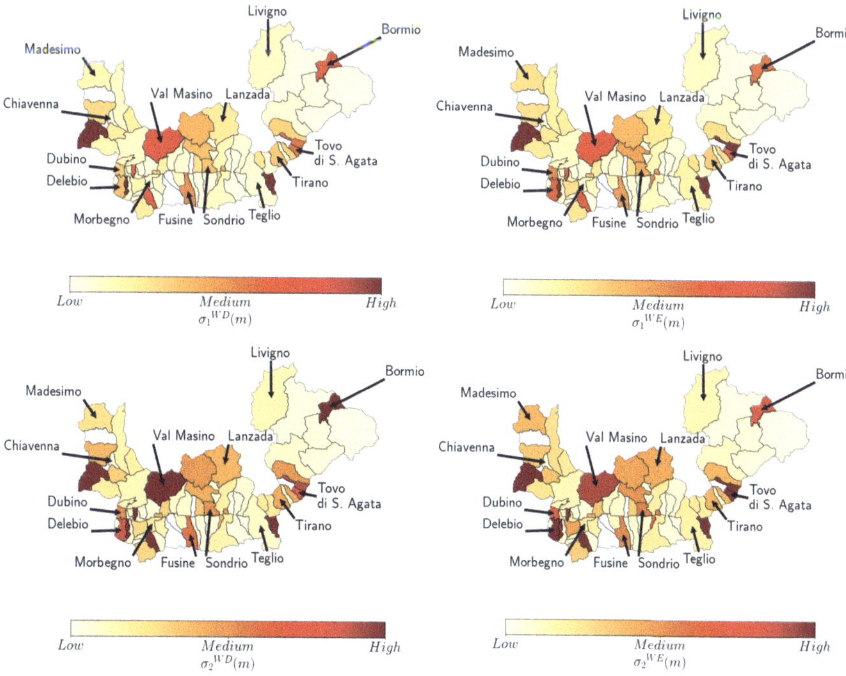

Fig. 7 Heatmaps of the Province of Sondrio. The color code represents the CHFP value of each municipality, with respect to workdays in $\boldsymbol{P_1}$ (upper left), weekends in $\boldsymbol{P_1}$ (upper right), workdays in $\boldsymbol{P_2}$ (lower left), and weekends in $\boldsymbol{P_2}$ (lower right). Source: image by the authors; data: MNO, ISTAT, and Data Portal of the Region of Lombardy, elaborated by the authors

to the corresponding value of σ_1^{WD} (top left), σ_1^{WE} (top right) σ_2^{WD} (bottom left), or σ_2^{WE} (bottom right). According to the index, we consider the cultural heritage fruition potential as *Low* when its reading is below 7.5 users per day, with a *High* potential being associated to CHFP values higher than 22.5 users/day. Finally, we associate CHFP values ranging between 7.5 and 22.5 users/day to *Medium* cultural heritage fruition potential. Note that the black lines in each map identify the administrative boundaries of each municipality within the Province, while black arrows are used to pinpoint some of them by their own names.

In Table 1, we report the municipalities boasting a *High* average cultural heritage fruition potential during P_1 (namely, σ_1) and P_2 (namely, σ_2), where the average is computed over work (i.e., σ_1^{WD} or σ_2^{WD}) and weekend days (i.e., σ_1^{WE} or σ_2^{WE}). Overall, they host roughly 43% of the cultural heritage sites in Valtellina: focusing on period P_2, *the heritage sites located in Bormio, Teglio, and Madesimo enjoy the potential visit of over 50 users, while those located in Tovo di S. Agata and Tirano get 40 to 50 potential visitors.*

On the one hand, a row-wise comparison of the heatmaps shows us how the distribution of CHFP values has changed from P_1 (plots on top row) to P_2 (plots on bottom row). As one can see, *the cultural heritage fruition potential of the whole Province has increased after the COVID-19 pandemic, when restrictions were repealed by the Italian government. Overall, CHFP has on average increased by 40%, a phenomenon mainly due to an increase in potential visitors during work rather than weekend days.* In fact, *while during workdays* (plots on left column) *the average increase equals 44%* (σ_1^{WD} = 6.7 versus σ_2^{WD} = 9.6 average users/day), *during the weekend* (plots on right column) *it decreases to 33%* (σ_1^{WE} = 8.7 versus σ_2^{WE} = 11.7 average users/day). To provide details on these aspects, we represent in Fig. 8 the average daily absolute increase (from P_1 to P_2) of potential visitors per cultural heritage site for the top-ten ranking municipalities.

Table 1 Average fruition potential of cultural heritage sites during $\boldsymbol{P_1}$ (i.e., $\boldsymbol{\sigma_1}$) and $\boldsymbol{P_2}$(i.e., $\boldsymbol{\sigma_2}$); the average herein is computed with respect to workdays and weekends. The table only includes municipalities with $\boldsymbol{\sigma_2}$>22.5. Municipalities are listed in a descending order with respect to $\boldsymbol{\sigma_2}$

m	σ_1 (Avg.)	σ_2 (Avg.)	$C(m)/\sum_m C(m)$ (%)
Bormio	78.3	90.1	14
Teglio	57.8	82.4	4.5
Madesimo	40.11	53.6	1.2
Tovo di S. Agata	22.53	48.9	0.8
Tirano	33.0	40.35	6.2
Ponte in Valt.	24.4	34.14	2.1
Valdidentro	15.0	32.5	2.0
Villa di Tirano	21.22	31.2	2.0
Delebio	23.1	29.7	1.8
Sondrio	20.0	27.9	8.4

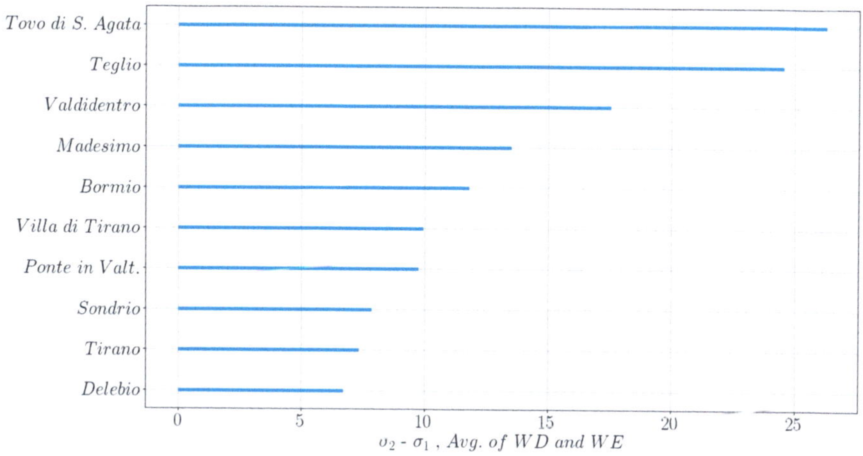

Fig. 8 Top-10 ranking of municipalities per average daily absolute increase (from P_1 to P_2) of potential visitors per cultural heritage site, in a decreasing sequence. Source: image by the authors; data: MNO, ISTAT, and Data Portal of the Region of Lombardy, elaborated by the authors

Table 2 Daily absolute and relative increase of average potential visitors per cultural heritage site from P_1 to P_2, where top-10 ranked municipalities are listed in a decreasing sequence with respect to the absolute increase

m	$\sigma_2 - \sigma_1$ (avg.)	$(\sigma_2 - \sigma_1)/\sigma_1$ (avg.)
Tovo di S. Agata	26.3	117%
Teglio	24.6	43%
Valdidentro	17.6	122%
Madesimo	13.5	41%
Bormio	11.8	17%
Villa di Tirano	9.9	47%
Ponte in Valtellina	9.7	40%
Sondrio	7.9	39%
Tirano	7.3	22%
Delebio	6.7	29%

The values of the increases are also reported in Table 2, along the corresponding relative increase observed in P_2 with respect to P_1. As one can see, *Tovo di S. Agata and Teglio saw an increase in their cultural heritage fruition potential by more than 24 users/day, with the former village more than doubling its potential after the Pandemic. The same trend is observed in Valdidentro, the CHFP of which increased by over 17 users/day. A relative increase between 40% and 48% can be observed in Villa di Tirano, Madesimo, and Ponte in Valtellina, although there the absolute increase is confined to 9–14 users daily. Relative increases lower than 40% are observed for the rest of municipalities in the top-10 ranking, among which Bormio is the one reporting the highest absolute increase equal to roughly 12 users/day.*

On the other hand, looking at the heatmaps by column lets one appreciate how the same distribution changes from workdays (left column) to weekends (right

Table 3 Average daily loss of potential visitors per cultural heritage site from work to weekend days, where top-10 ranked municipalities are listed in a decreasing order. The average is computed between the losses measured in $\boldsymbol{P_1}$ and $\boldsymbol{P_2}$

m	$C(m)/\sum_m C(m)$ (%)	$\sigma^{WE} - \sigma^{WD}$ (avg.)
Sondrio	8.4	**−5.9**
Tovo di S. Agata	0.8	**−4.5**
Delebio	1.8	**−3.2**
Albosaggia	0.9	−1.8
Morbegno	3.8	−1.5
Mazzo di Valtellina	2.2	−0.9
Grosotto	1.2	−0.8
Civo	1.2	−0.4
Spriana	0.1	−0.4
Cosio	2.0	−0.3

column). *Regardless of the period under consideration, most Valtellina municipalities are visited with an either equal or more higher likelihood during a weekend than a workday. However, while in* P_1 *the relative average increase in the CHFP from workdays to weekends equals 31%* (σ_1^{WD} = 6.70 versus σ_1^{WE} = 8.74 average users/day), *this gap decreases to 21% in* P_2 (σ_2^{WD} = 9.60 versus σ_2^{WE} = 11.65 average users/day). In fact, *the average fraction of cultural heritage fruition potential due to workday visitors has raised from 65.6% (*P_1*) to 67.3% (*P_2*). In other words, Valtellina attracted, on average, more users during workdays compared to the time preceding the pandemic, a likely result of the many investments devoted by public entities to upgrade the digital infrastructures throughout the Valley. We believe that such efforts will greatly improve visitor experience, especially thanks to enhanced remote working options. Nevertheless, we ought to underline that such trend is not observed for all municipalities.* To illustrate this aspect, we report in Table 3 the top-10 average losers in the number of potential weekend visitors, i.e., the average gap between σ^{WD} and σ^{WE} with respect to $\boldsymbol{P_1}$ and $\boldsymbol{P_2}$. As one can see, *Sondrio, Delebio, and Tovo di S. Agata are on top of the ranking, missing during the weekend an average of 5.2, 2.2, and 1.8 users/day with respect to workdays, respectively.*

These results let us provide an answer to question *Q2*: *according to the proposed metrics, high average fruition potential in period* P_2 *is associated to Bormio (90 visitors/day), Teglio (82 visitors/day), Madesimo (53 visitors/day), Tovo di S. Agata (48 visitors/day), and Tirano (40 visitors/day). Overall, we observe a 40% positive variation of CHFP in Valtellina after the pandemic, and such variation depends on the municipality being considered. Higher variations (on average, more than 10 potential visitors/day) are observed in Tovo di S. Agata, Teglio, Valdidentro, Madesimo, and Bormio. Also, while we observe on the one hand that municipalities in Valtellina get more potentially visited during weekends than on workdays, on the other hand the post-pandemic positive variation of such potential is 11% higher during workdays than weekends: some exceptions to this trend are represented by Sondrio, Tovo di S. Agata, and Delebio.*

Limitations and Future Research Developments

Let us now highlight some limitations to our work, which in turn warrant some future research on our part:

1. Firstly, we solely set our sights on Sondrio, an individual province characterized by a relatively specific orography, population density and distribution, and available infrastructures including the provisioning of Internet endpoints to local users. We thus plan to benchmark our results considering a wider array of scenarios to include both inland and metropolitan areas.
2. Secondly, we would like to stress that the mobile radio data considered provide no information as to the reasons why any given user would be spending time in Valtellina during the periods under consideration. Some insight as to this latter aspect might be provided by sourcing cellular user mobility data (e.g., Origin-Destination matrices), rather than performance KPIs measured as aggregates at the access of the network.
3. Thirdly, we approximated the coverage of cellular base stations using a Voronoi tessellation of the Province of Sondrio. A more accurate representation of the coverage provided by a set of cellular base stations would require the mobile network operator to specify the mapping between physical locations and most likely serving radio access point, which is generally achieved by using ad-hoc radio coverage maps.
4. Finally, we based our estimate of the number of cellular users connected from a given municipality on the assumption that the urban footprint of such municipality is uniformly distributed within its administrative area. Indeed, such footprint is often more densely concentrated in a limited area within the municipality's administrative borders, especially in the case of inland or rural scenarios. In such cases, we should rule out as valid sources of data all intersections between the coverage area of a cellular base station and the administrative area of a municipality when they do not include any portion of the municipality's urban footprint. Furthermore, and even in the case of a urban footprint uniformly spreading within a given administrative area, Point of Interest density unlikely follows suit, thus further biasing the final estimate. In future works, we plan to address this issue by considering the exact shape of each municipality urban footprint, e.g. resorting to information provided by CORINE Land Cover database.[19]

[19] https://land.copernicus.eu/en/products/corine-land-cover.

Conclusions

Undoubtedly, the information regarding the presence and mobility patterns of people in a city or town, as well as the density of their activity in different areas are of paramount importance to urban planners, designers, cultural heritage managers, policy makers, etc. On the one hand, many improvements in the comprehension of user dynamics thanks to the knowledge extracted from cellular data have enabled novel applications in the context of smart cities, such as mobile traffic projections, transportation design, and virus-spread modelling. On the other hand, the identification of areas with denser human activity as well as the analysis of its evolution over time provide some crucial insight as to where and when a specific action is mostly needed, thus driving decisions and investment by policy makers and institutions.

In the cultural heritage sector, results derived from mobile data analysis may well provide some relevant insight as to which buildings/areas are over- or under-attended. This type of study can support cultural heritage management activities, such as: measuring overall interest in a specific cultural heritage site, understanding whether any protection measure is required on the basis of carrying capacity, identifying sites that deserve improved valorization activities, pursuing solutions for improved tourist distribution, addressing well thought-through policies, etc.

The pandemic drew attention to the valorization of inland area heritage and resulted in changes that need to be monitored. In the case study presented herein, mobile data analysis pointed out the following key aspects to be taken into account by local stakeholders (policy makers, cultural heritage managers, consortiums, mountain communities, chambers of commerce, local business owners, wine and agricultural companies, etc.):

- The number of mobile users within the Province of Sondrio increased by an average of 30% after the pandemic. Besides, the fruition potential of cultural heritage sites increased by an average of 40%. This may serve as confirmation that the pandemic has revived the overall interest towards inland areas, thus attracting more people and increasing opportunities for the local economy, to be triggered by the tangible and intangible resources typically characterizing these areas.
- Considering that the number of users tends to increase at weekends, some action should be taken to intercept different target groups during workdays, such as students, retirees, remote workers, etc. Indeed, we ought to stress that in the years following the pandemic the relative number of potential heritage visitors has increased mainly on weekdays, likely following the spread of remote working.
- Since seasonal tourism related to winter sports and spas still dominates in the upper Valley, we may have to take into account operations aimed at de-seasonalizing tourism and intercepting at a mid-valley level all tourists heading uphill for spas and mountain sport purposes. Diversifying the tourism offer by adding to it novel services and opportunities such as strengthening the summer

sports offer, increasing the promotion of the Route of Terraces, or introducing new cultural initiatives might well prove conducive to this objective.
- Some small, lesser-known towns, such as Pedesina and Madesimo, have a high density of cultural heritage sites, which probably deserve to be better valorized.
- Given that other small towns such as Tovo di Sant'Agata and Teglio experienced a doubling in their number of potential heritage users after the pandemic, it might be of use to investigate the reasons why this significant change even took place. Valdidentro, Villa di Tirano, Madesimo, and Ponte in Valtellina, too, almost doubled their user numbers.

To conclude, we should underline that the results presented in this paper are the first step of an attempt to use mobile radio data to address strategies for the valorization of cultural heritage.

Further investigations on this topic are warranted to provide novel insight thanks to this type of analysis. In particular, future research vectors shall focus on aspects related to (1) the identification of users mobility flows trajectories and how they have changed over time (e.g., before and after the pandemic) and (2) how digital infrastructure use (either fixed, mobile, or both) correlates with application types, e.g., how interactions with museum or tourist attraction-related web-sites have changed over time.

Acknowledgements The authors convey their sincere gratitude to Prof. Giovanna Fossa of the Department of Architecture, Built Environment, and Construction Engineering of Politecnico di Milano and Prof. Brent D. Ryan of the Department of Urban Studies and Planning of MIT for the opportunity of taking part in this research project. Special thanks, also, to Prof. Antonio Capone and Prof. Stefano Della Torre for supervising all current research activity.

References

AGCOM. (2024). *Annual report to the Italian parliament.* Retrieved October 31, 2024, from https://www.agcom.it/sites/default/files/documenti/relazione_annuale/RELAZIONE%20ANNUALE_2024.pdf

Avrami, E. (2019). Heritage data and the next generation of preservation policy. In E. Avrami (Ed.), *Preservation and the new data landscape, Columbia books on architecture and the city, series: Issues in preservation policy*. Columbia University Press.

Ayan, N., Damasceno, N. L., Chaskar, S., de Sousa, P. R., Ramesh, A., Seetharam, A., & de Rocha, A. A. (2021). Characterizing human mobility patterns during COVID-19 using cellular network data. In L. Khoukhi, S. Oteafy, & E. Bulut (Eds.), *Proceedings of the IEEE 46th conference on local computer networks (LCN 2021)* (pp. 471–478). IEEE.

Barbetta, G. P., Cammelli, M., & Della Torre, S. (Eds.). (2013). Distretti culturali: dalla teoria alla pratica. Il Mulino.

Barca, F., & Lucatelli, S. (2014). Introduzione. In F. Barca, P. Casavola, & S. Lucatelli (Eds.), *Strategia Nazionale per le Aree Interne, Collana Materiali UVAL (31)* (pp. 7–9). Agenzia per la Coesione Territoriale.

Bellini, A. (1996). A proposito di alcuni equivoci sulla conservazione. *TeMa, 1*, 2–3.

Benetti, D., & Guidetti, M. (1990). *Storia di Valtellina e Valchiavenna. Una introduzione.* Jaka Book.

Bonfantini, B. (Ed.). (2016). *Attivare risorse latenti. Metodi sperimentali per l'analisi, la mappatura e la gestione informativa integrata delle trasformazioni di territori e manufatti del patrimonio culturale diffuso*. Planum Publisher.

CHCfE Consortium. (2015). *Cultural heritage counts for Europe, full report*. Retrieved October 31, 2024, from http://www.encatc.org/culturalheritagecountsforeurope/outcomes/

Della Torre, S. (1999). "Manutenzione" o "Conservazione"? La sfida del passaggio dall'equilibrio al divenire. In G. Biscontin & G. Driussi (Eds.), *Ripensare alla manutenzione: ricerche, progettazione, materiali, tecniche per la cura del costruito, Atti del XV Convegno Scienza e Beni Culturali Bressanone* (pp. 71–80). Arcadia Ricerche.

Della Torre, S. (2015). Shaping tools for built heritage conservation: From architectural design to program and management. Learning from "Distretti culturali". In K. Van Balen & A. Vandesande (Eds.), *Community involvement in heritage* (pp. 93–101). Garant.

Ericcson. (2024). *Ericcson mobility report June 2024*. Retrieved October 31, 2024, from https://www.ericsson.com/en/reports-and-papers/mobility-report/reports/june-2024

Europa Nostra. (2020). COVID-19 & beyond. Challenges and opportunities for cultural heritage. Retrieved October 31, 2024, from https://www.europanostra.org/wp-content/uploads/2020/10/20201014_COVID19_Consultation-Paper_EN.pdf

European Commission. (2020a). *Recovery plan for Europe*. Retrieved October 31, 2024, from https://commission.europa.eu/strategy-and-policy/recovery-plan-europe_en

European Commission. (2020b). *The 2021–2027 EU budget – What's new? The main novelties of the EU's long-term budget, its revenue and spending areas and the NextGenerationEU recovery instrument*. Retrieved October 31, 2024, from https://commission.europa.eu/strategy-and-policy/eu-budget/long-term-eu-budget/2021-2027/whats-new_en

European Commission, Directorate-General for Budget. (2020). *The EU budget powering the recovery plan for Europe, publications office*. Retrieved October 31, 2024, from https://op.europa.eu/en/publication-detail/-/publication/e0956910-a0c9-11ea-9d2d-01aa75ed71a1/language-en

European Commission, Directorate-General for Budget. (2021). *The EU's 2021–2027 long-term budget & NextGenerationEU: Facts and figures, Publications Office*. Retrieved October 31, 2024, from https://op.europa.eu/en/publication-detail/-/publication/d3e77637-a963-11eb-9585-01aa75ed71a1/language-en

European Commission, Joint Research Centre. (2022). *New European Bauhaus. About the initiative*. Retrieved October 31, 2024, from https://new-european-bauhaus.europa.eu/about/about-initiative_en

European Heritage Alliance Manifesto. (2020). *Cultural heritage: A powerful catalyst for the future of Europe*. Retrieved October 31, 2024, from https://www.europanostra.org/wp-content/uploads/2020/09/20200915-EHA-MANIFESTO-layout_high-res.pdf

Foppoli, D. (2018). Cultural heritage as a resource for regional sustainable development: The example of the Valtellina Cultural District in Italy. In K. Van Balen & A. Vandesande (Eds.), *Innovative built heritage models, reflections on cultural heritage theories and practice* (Vol. 3, pp. 167–174). Taylor & Francis Group.

Foppoli, D., & Di Capita, F. (2018). The route of terraces in Valtellina: Community involvement and tourism for the enhancement of cultural landscape. In F. Alberti, A. Dal Pozzo, D. Murtas, M. A. Salas, & T. Tillmann (Eds.), *Paesaggi terrazzati: scelte per il futuro. Terzo incontro mondiale. Terraced landscapes: Choosing the future. Third world meeting* (pp. 401–408). Regione del Veneto, Direzione Pianificazione Territoriale, Strategica e Cartografia.

Foppoli, D., Konsta, A., & Della Torre, S. (2019). Education and training as a means to boost skills and to increase territory reputation: The example of Valtellina Cultural District in Italy. In K. Van Balen & A. Vandesande (Eds.), *Professionalism in the built heritage sector, reflections on cultural heritage theories and practice* (Vol. 4, pp. 95–102). Taylor & Francis Group.

Interreg Italia-Svizzera. (2024). Retrieved October 31, 2024, from https://www.interreg-italiasvizzera.eu/wps/portal/site/interreg-italia-svizzera/DettaglioRedazionale/progetti-2014-2020/convalore

KEA European Affairs. (2020). *The impact of the COVID-19 pandemic on the cultural and creative sector*. Report for the Council of Europe. Retrieved October 31, 2024, from https://keanet.eu/wp-content/uploads/Impact-of-COVID-19-pandemic-on-CCS_COE-KEA_26062020.pdf.pdf

Lanza, G., Pucci, P., Carboni, L., & Vendemmia, B. (2022). Impacts of the Covid-19 pandemic in inner areas. Remote work and near-home tourism through mobile phone data in Piacenza Apennine. *TEMA Journal of Land Use, Mobility and Environment, 2*, 73–89.

Legeby, A., Koch, D., Duarte, F., Heine, C., Benson, T., Fugiglando, U., & Ratti, C. (2023). New urban habits in Stockholm following COVID-19. *Urban Studies, 60*(8), 1448–1464.

Lynch, K. (1972). *What time is this place?* The MIT Press.

Mariotti, V. (Ed.). (2016). *Dinamiche insediative nelle Alpi centrali tra antichità e Medioevo*. SAP Società Archeologica.

Martinez-Durive, O. E., Couturieux, T., Ziemlicki, C., & Fiore, M. (2022). VoronoiBoost: Data-driven probabilistic spatial mapping of mobile network metadata. In *19th annual IEEE international conference on sensing, communication, and networking (SECON)* (pp. 100–108). IEEE. https://doi.org/10.1109/SECON55815.2022.9918610

Maucorps, A., Römisch, R., Schwab, T., & Vujanovic, N. (2022). The future of EU cohesion. In *Effects of the twin transition on disparities across European regions*. Bertelsmann Stiftung. https://doi.org/10.11586/2022127

Ministero della Cultura. (2021a). *Avviso pubblico progetti di rigenerazione culturale e sociale dei piccoli borghi storici PNRR M1C3 – investimento 2.1 – attrattività dei borghi – linea B*. Retrieved October 31, 2024, from https://cultura.gov.it/borghi

Ministero della Cultura. (2021b). *PNRR, Franceschini: il Piano Nazionale Borghi per vincere la sfida del ripopolamento*. Retrieved October 31, 2024, from https://cultura.gov.it/comunicato/21911#allegati

Ministero della Cultura. (2021c). *Cultura. Next generation Eu. Recovery and resilience plan, 2021*. Retrieved October 31, 2024, from https://media.beniculturali.it/mibac/files/boards/be78e33bc8ca0c99bff70aa174035096/PDF/Il%20Piano%20Nazionale%20di%20Ripresa%20e%20Resilienza%20(PNRR).pdf

Muench, S., Stoermer, E., Jensen, K., Asikainen, T., Salvi, M., & Scapolo, F. (2022). *Towards a green and digital future, EUR 31075 EN, JRC129319*. Publications Office of the European Union. https://doi.org/10.2760/977331

Naboulsi, D., Fiore, M., Ribot, S., & Stanica, R. (2015). Large-scale mobile traffic analysis: A survey. *IEEE Communications Surveys & Tutorials, 18*(1), 124–161. https://doi.org/10.1109/COMST.2015.2491361

OECD. (2001). *Understanding the digital divide*. Retrieved October 31, 2024, from https://www.oecd.org/en/publications/understanding-the-digital-divide_236405667766.html

Ookla. (2023). *The state of Worldwide connectivity in 2023*. Retrieved October 31, 2024, from https://www.ookla.com/articles/worldwide-connectivity-mobile-fixed-networks-digital-divide-2023

Ookla. (2024). *Speedtest global index*. Retrieved October 31, 2024, from https://www.speedtest.net/global-index#mobile

Paesaggi della Media Valtellina. (2023). Retrieved December 22, 2022, from https://www.paesaggidivaltellina.it

Pimpinella, A., Boniotti, C., Ignaccolo, C., Martignon, F., Pavon, A., & Venturini, L. (2024). Data-driven profiling of inland areas: Studying changes in mobile users presence after COVID-19. In *IEEE international symposium on personal, indoor and mobile radio communications (PIMRC)*. IEEE.

Pimpinella, A., Di Giusto, F., Redondi, A. E. C., Venturini, L., & Pavon, A. (2022). Forecasting busy-hour downlink traffic in cellular networks. In *IEEE international conference on communications (ICC)* (pp. 4336–4341). IEEE. https://doi.org/10.1109/ICC45855.2022.9838982

Più Segni Positivi. (2022). Retrieved August 29, 2022, from www.piusegnipositivi.org

Potts, A. (2021). *European cultural heritage green paper*. Europa Nostra.

Ryan, B. D. (2017). *The largest art. A measured manifesto for a plural urbanism*. The MIT Press.

Sandvine. (2023). *The global internet phenomena report*. Retrieved October 31, 2024, from https://www.sandvine.com/hubfs/Sandvine_Redesign_2019/Downloads/2023/reports/Sandvine%20GIPR%202023.pdf

Singh, R., Fiore, M., Marina, M., Tarable, A., & Nordio, A. (2019). Urban vibes and rural charms: Analysis of geographic diversity in mobile service usage at national scale. In L. Liu & R. White (Eds.), *WWW'19: The world wide web conference* (pp. 1724–1734). Association for Computing Machinery.

Tantillo, F. (2019). Quattro anni di Strategia Nazionale Aree Interne. Il Giornale delle Fondazioni, Speciale Montagna. Risorsa del XXI secolo. Un percorso di riflessioni e ascolti sulla montagna e le aree interne (pp. 52–54).

Territorial Agenda 2030. (2020). *A future for all places. Informal meeting of Ministers responsible for spatial planning and territorial development and/or territorial cohesion*. Retrieved October 31, 2024, from https://territorialagenda.eu/wp-content/uploads/TA2030_jun2021_en.pdf

UNESCO – United Nations Educational, Scientific and Cultural Organization. (2016). *Culture urban future. Global report on culture for sustainable urban development*. Retrieved October 31, 2024, from http://openarchive.icomos.org/id/eprint/1816/1/245999e.pdf

Villa, D. (2017). Open data for cultural heritage. In *Place based digital humanities between representation, design and innovation*. Planum Publisher.

Vitale, C. (Ed.). (2024). *Innovazione ed inclusione per la valorizzazione del patrimonio culturale e lo sviluppo delle aree interne. Idee e proposte*. Giappichelli Editore.

Vitale, M. (2019). Introduzione. In D. Benetti, I luoghi del vino di Valtellina. *Cooperativa editoriale Quaderni valtellinesi*.

Wallace, M., Poulopoulos, V., Antoniou, A., & López-Nores, M. (Eds.). (2023). *Big data analytics for cultural heritage*. MDPI.

Work from Home. (2024). Retrieved October 31, 2024, from https://wfh-inequalities.netlify.app/

A Digital Platform for Italian Small Businesses

Jonathan Goh and Juanita Halim

Introduction

Italian Small and Medium-Sized Enterprises (SMEs) play a pivotal role in the Italian economy. There are about 4.3 million SMEs in Italy, 95% of which are microenterprises. They account for 80% employment and 66.9% of value added (European Investment Bank, 2021). SMEs contribute to 53% of the country's total exports, higher than EU average for about 40%, and their closest neighboring countries, Germany and France for about 23%.[1] Many Italian SMEs drive the economy by manufacturing high-quality consumer goods and cultural art pieces–of which private family companies produce the majority. Some of these businesses are passed down through multiple generations. These types of businesses add intangible cultural heritage that form national and regional identity and help drive the tourism economy. Tourism generated by these SMEs benefit communities directly in the form of revenue generated from services and commercial articles through hotels, transportation, and traditional craft goods, and indirectly by receiving improvements in infrastructure, urban rehabilitation, conservation of certain historical sites and cultural heritage.

[1] Data source: Eurostat (2017).

J. Goh
Oracle, Austin, TX, USA

MIT Masters of City Planning 2023, Cambridge, MA, USA
e-mail: j_goh@mit.edu

J. Halim (✉)
MIT Masters of City Planning 2023, Cambridge, MA, USA
e-mail: jhalim@alum.mit.edu

G. Fossa et al. (eds.), *Small-Town Renaissance*, Research for Development,
https://doi.org/10.1007/978-3-031-89786-3_3

However, these SMEs alongside the Italian economy face decline cause by rapid depopulation Italy's economic model is very dependent on family-owned companies that are typically smaller and less productive than their equivalents elsewhere. This problem has grown worse in the recent decades. Younger generations that are more affluent with technology are migrating to bigger cities or neighboring countries such as Milan or Switzerland. The loss of high-skilled workers in Italian shrinking villages negatively affects the local innovation of SMEs, disconnecting them with the global economy (ESPON, 2017), especially after the COVID-19 pandemic hits. This hits small and medium-sized enterprises (SMEs) and micro-enterprises hardest as they are more dependent on the availability of skilled worker (Tietjen & Jørgensen, 2016). Despite there are numbers of small businesses in Italy already transitioning to the digital world (e.g., cloud), their lag in adopting e-commerce is an obstacle in this situation.

Moreover, data from the European Commission Report 2021 stated that SMEs in Italy find it more difficult to access staffing resources with appropriate digital media compared to their EU peers. Italian SMEs are penalized by a larger proportion of the population still lacking basic digital skills, translating into a lower level of information and communication technology (ICT) graduates (1.4% of Italy graduates versus EU average of 3.9%, France and Germany of 3.9%) (European Commission, 2022). Most Italian small and medium enterprises (69%) have at least a basic level of digital intensity, a share that is well above the EU average (60%). Italian enterprises perform very well in the use of e-invoices, although gaps remain in the use of technologies such as big data and Artificial Intelligence, and in the uptake of e-commerce (European Commission, 2022). This makes SMEs in Italian small towns suffer a lot more, especially since the COVID-19 pandemic.

To overcome delays and bridge the gaps between Italy and other EU countries, it is essential to consistently focus on a comprehensive strategy that encompasses policies on human capital, innovation, and business competitiveness. By effectively implementing the initiatives already undertaken and leveraging the measured outline in the Italy's Recovery and Resilience Plan (RRP), Italy can bring about a substantial change and seize the opportunity to promote digitalization throughout the nation.

Digital Divide Between SMEs

There is a large digital divide between SMEs in the northern and southern region of Italy; however, even Italy's best performing regions have significant gap versus the rest of Europe. Italy ranks 18th out of 28 EU Member States in the 2022 edition of Digital Economy and Society Index (DESI). As the third largest EU economy, Italy's progress in the digital transformation over the coming years is crucial to enable the EU as a whole to reach the 2030 Digital Decade targets (European Commission, 2022).

Data from DESI 2022 shows that generally seven out of nine regions with a score above the Italian average are from the north: Lombardy, Emilia-Romagna, Liguria, Piedmont, Veneto, and the autonomous provinces of Trento and Bolzano. In particular, the lowest are regions in the south (Balocco et al., 2019).

Data and Methodology

Hourly cellphone data, economic data, demographic data, and small business owners' interviews are keys to this project analysis and outcome. For this project, we have applied a multi-method quantitative and qualitative approach on (1) site visits, (2) comprehensive data analysis of hourly cellphone activity data, economic and demographic data, and (3) interviews with key small business owners from sample villages in Lombardy and Sicily.

We began the project in August 2022 with introductory site visits and interviews with local governments and community actors in 7 Italian shrinking villages across Lombardy (Sondrio, Teglio, and Tirano) and Sicily (Favara, Mazzarino, Vizzini, and Centuripe). Following the site visits, we decided to study economic and demographic data from the villages we visited and thus, provided us with detailed insight into selecting the pilot area for the project. In practice, we studied the available open data from ISTAT, Eurostat, Geoportale Lombardia, Opendata Regione Sicilia, and hourly cellphone data activity from the years 2020 and 2022.

After analyzing the economic and demographic, we conducted interviews with 10 small business owners: 6 from Sondrio, 3 from Centuripe, and 1 from Vizzini, to understand what problems they are currently facing and how digitalization can help them thrive (Fig. 1). We interviewed 10 small business owners from different sectors: food, art, agriculture, hospitality, food service, and agritourism, to understand their business landscape and scope, typical customers, community, and current digital platform adoption.

Findings

From the demographic and economic data, we found that the number of small businesses/artisan enterprises in Lombardy and Sicilia has continuously declined for the past decades (Figs. 2 and 3), with Sicily having fewer than Lombardy. However, it is interesting to see that number of businesses selling online is higher in Sicily than in Lombardy despite Lombardy having slightly higher businesses using websites in the Lombardy region (Fig. 4).

From the interviews with small business owners, while business owners use the website for some aspect of their business, the level of digital adoption between businesses varies. Some businesses have used digital platforms like websites or social media to promote businesses. In contrast, some others have integrated blockchain

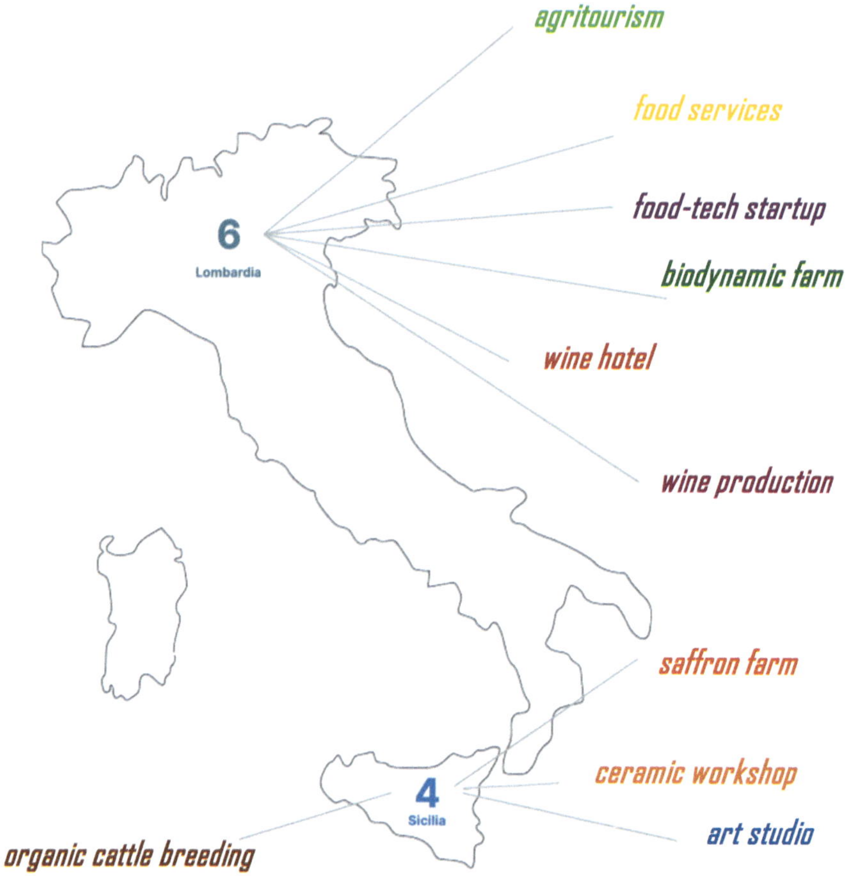

Fig. 1 Number of SMEs interviewed in Lombardy and Sicily region. Source: image by the authors

into their system and developed their in-house mobile and web applications. We learned that businesses in Sondrio, Lombardy region, have a higher level of digital adoption for their businesses.

From our site visits and interviews, we found several distinct key characteristics. One notable aspect is their strong emphasis on fostering local relationships. These SMEs recognize the importance of building connections with their communities. This results in a mutually beneficial ecosystem where businesses support one another, share resources and create a sense of solidarity to envision a vibrant local economy. Furthermore, the region's rich cultural history plays a significant role in shaping SMEs in the region. These enterprises often draw inspiration from centuries-old traditions, craftsmanship, and artisanal skills, integrating them into their operations. This cultural heritage adds a distinct flavor to their products and services, attracting domestic and international customers seeking authenticity and a connection to the past.

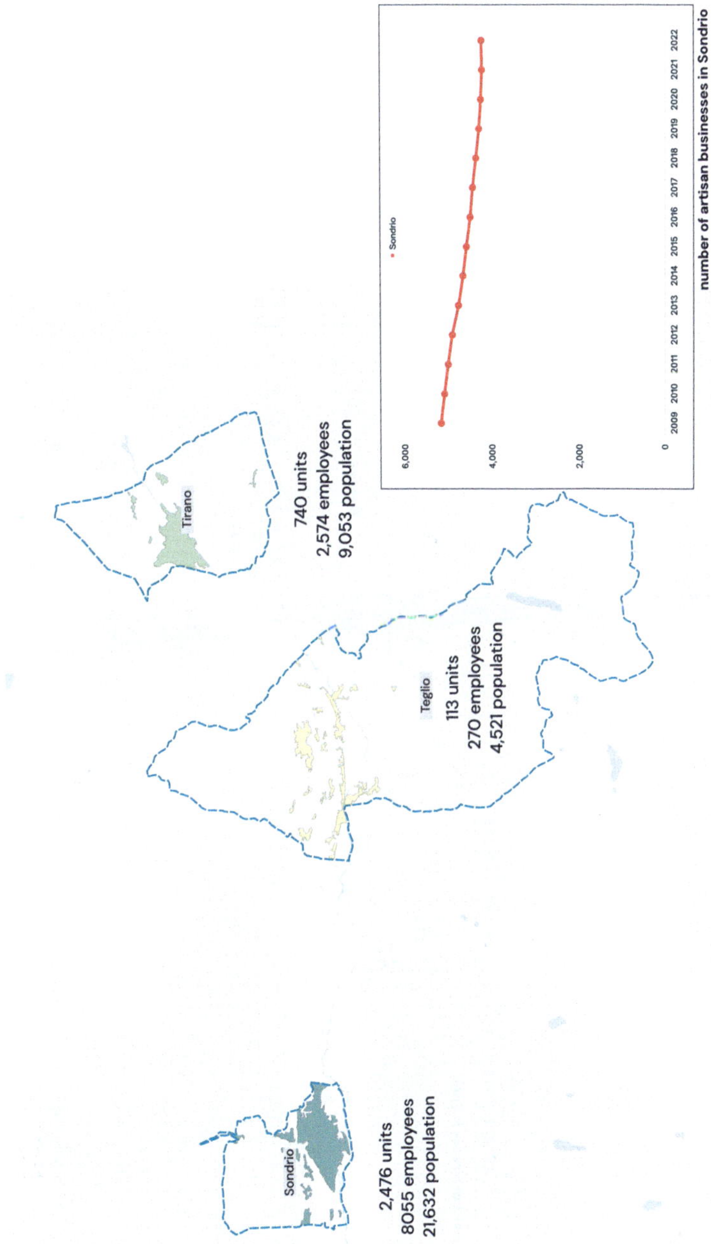

Fig. 2 Number of small/artisan businesses in Sondrio province, Lombardy. Source: image by the authors; data: ISTAT

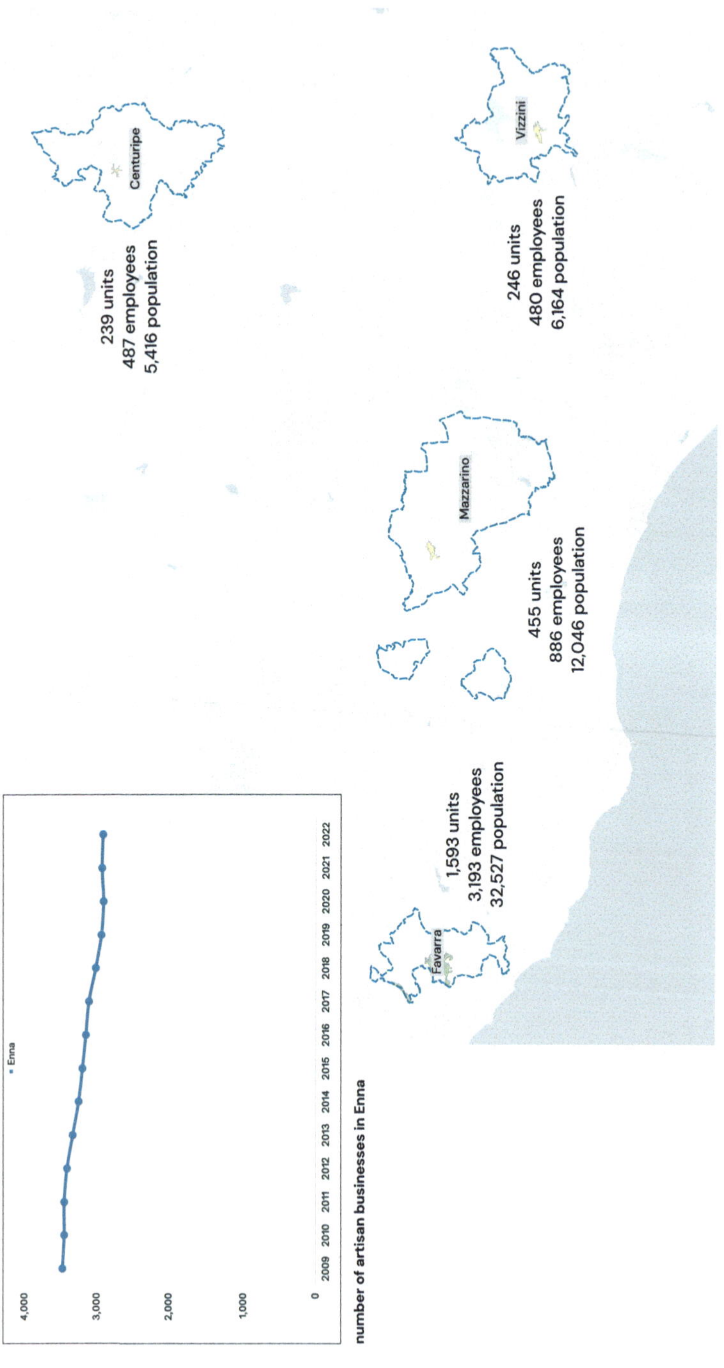

Fig. 3 Number of small/artisan businesses in Enna province, Sicily region. Source: image by the authors; data: ISTAT

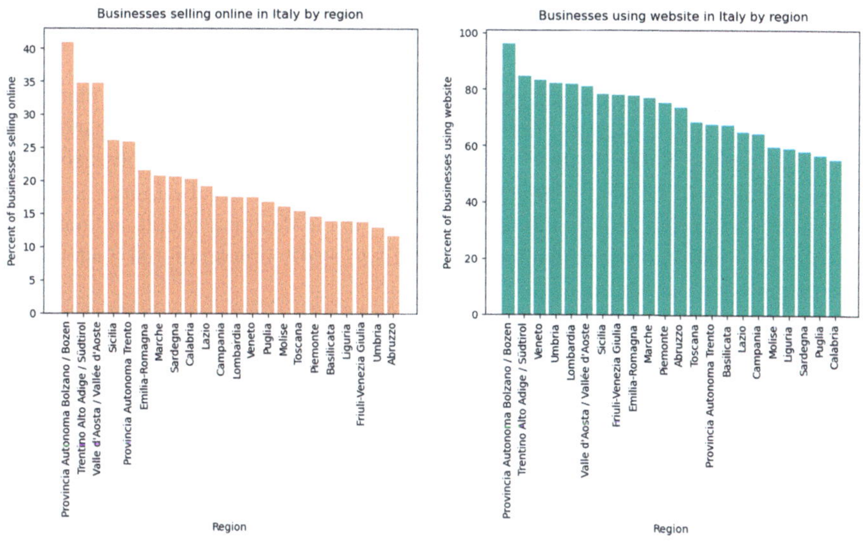

Fig. 4 Businesses selling online and using website in Italy by region. Source: image by the authors; data: European Investment Bank (2021)

Another key characteristic of Italian SMEs is their specialization in producing unique, high-quality goods. They excel in crafting niche products that embody excellence and craftsmanship, whether it be fashion, design, food, or other industries. These specialized products often carry a sense of exclusivity. Consequently, they are highly sought after, catering to discerning customers who appreciate the artistry and attention to detail involved.

However, while Small and Medium Enterprises (SMEs) in Italy boast a unique blend of characteristics contributing to their success, there are still opportunities for improvement. One area for improvement is marketing, specifically in customer acquisition, awareness, engagement, and conversion. SMEs must adopt modern marketing strategies that leverage digital channels to reach a wider audience and engage customers more effectively.

A notable avenue for SMEs involves engaging in collaborations with fellow business owners, fostering the development of a vibrant local economy. By working together, SMEs can pool resources, share knowledge, and significantly impact their communities. This collaborative approach can help SMEs overcome challenges, such as limited resources and competition from larger businesses, and foster a more resilient and sustainable local economy.

Finally, SMEs in Italy must develop their place identity by creating in-person and virtual experiences that showcase their unique culture, history, and products. By creating memorable customer experiences, SMEs can differentiate themselves from competitors, attract new customers, and build a loyal fan base. In turn, this can help SMEs increase their revenue, boost their brand reputation, and contribute to the overall economic development of their regions.

Design Concept

Guided by our user research, we propose a place-based, multi-sided web platform that promotes regional cultural heritage by integrating digital storytelling and collaboration among SMEs in Italy. Our first target users—geographically isolated SMEs in Italy will be able to use the platform to promote local cultural heritage experiences and products. On the back end, we plan to design a simple onboarding and content creation process that enables SMEs to easily upload digital content for their business on the platform.

We learned from our conversations with SME owners that they currently have no easy way of working together with other SMEs in this area, despite the fact that their relationships with other SMEs are vital to their businesses' success. Hence, one of the key features of the platform is to make it easy for SMEs to meet, collaborate, and build community with a network of other SMEs.

The second target user of the platform are potential visitors to less-frequent visited part of Italy (Fig. 5). The platform offers potential visitors a way to connect with Italian SMEs in small towns and learn more deeply about the area. This more personal connection with a sense of place combined with trip-planning features through the platform will make the experience of visiting small towns more seamless.

Features

Distinguished by its unique attributes, our platform stands apart from conventional e-commerce platform. We are dedicated to simplifying the digitization process for small businesses, offering them as seamless pathway to establish a robust online presence that helps attract local and international tourists to visit.

Interactive Place-Based Storytelling

Our platform is built around the compelling art of storytelling, enabling businesses to forge intimate connections with locales and their rich heritage, thus enhancing the personal bond between customers and places. Using an existing 360 camera model of an existing local cultural heritage, we engage local cultural heritage and immersive narratives rooted in specific locations, where potential customers could explore digitally, shaping a profound and interactive experience that brings local stories to life (Fig. 6).

Fig. 5 Simulation of the proposed web platform selling SME-driven local experiences and products. Source: image by the authors; basemap: Google Earth Image 2025 Airbus

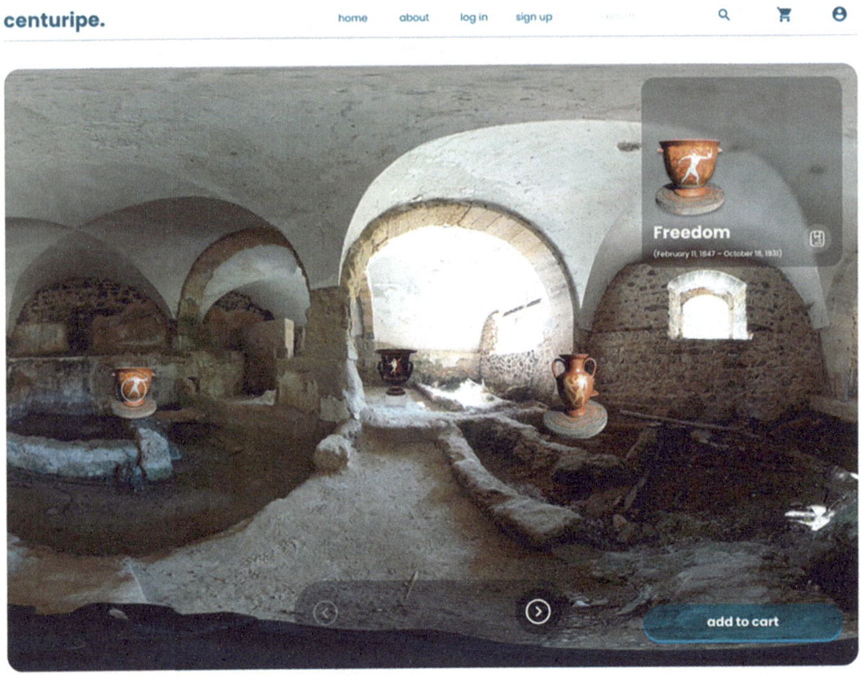

Fig. 6 Simulation of the proposed web platform: interactive place-based storytelling using a 360-model of a cultural heritage building in Vizzini. Source: image by the authors

Interactive Place-Based Exploration

Our platform introduces an innovative approach to interactive place-based exploration, offering users an immersive journey through the heart of Italian small towns and their intricate tapestry of SMEs. Through cutting-edge technology, users can embark on virtual expeditions that seamlessly blend cultural heritage, local experiences, and economic vitality. This interactive exploration provides a window into the distinctive charm and character of each town, fostering a deeper understanding of its history, traditions, and the integral role played by SMEs in shaping its identity (Fig. 7).

Direct Connection with SMEs

Our platform establishes a direct and meaningful connection between users and small business owners, bridging the gap between consumers and the heart of local economies. By providing a seamless avenue for communication, users can engage directly with passionate entrepreneurs who are the driving force behind Italian small

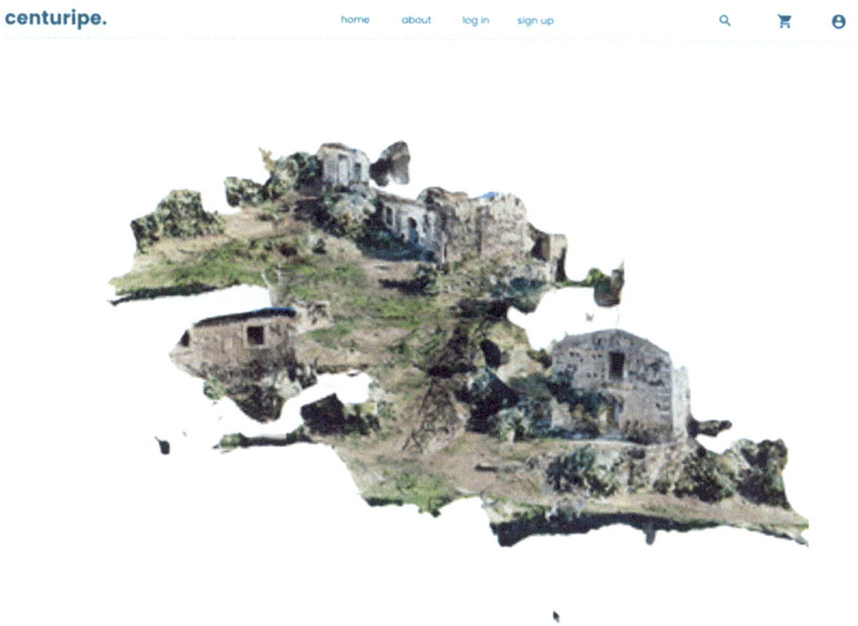

Fig. 7 Simulation of the proposed web platform: interactive place-based exploration. Source: image by the authors elaborated through cutting-edge technologies

towns' economic vibrancy. This direct connection empowers users to gain firsthand insights, explore unique products and services, and even participate in collaborative initiatives, fostering a sense of community and shared growth (Fig. 8).

Collaborative Experience Development

Our platform's ethos lies in the synergy between collaborative experience development and the concept of *territorial capital (Perucca, Eur Plan Stud 22(3):537–562, 2014)*. By embracing the notion of *territorial capital*, we acknowledge the intrinsic value of local connections and relationships within Italian small towns. Through collaborative efforts, SMEs harness their collective strengths, weaving a rich tapestry of experiences that reflect the authentic spirit of each community. This collaboration not only magnifies the economic impact but also nurtures a sense of pride and shared purpose among small business owners. As these enterprises contribute to the growth of their local economies, they cultivate a reservoir of cultural and social capital that further enriches the fabric of the community. In turn, our platform becomes a conduit for fostering these connections, promoting economic vitality, and amplifying the unique identity of Italian small towns on a global stage.

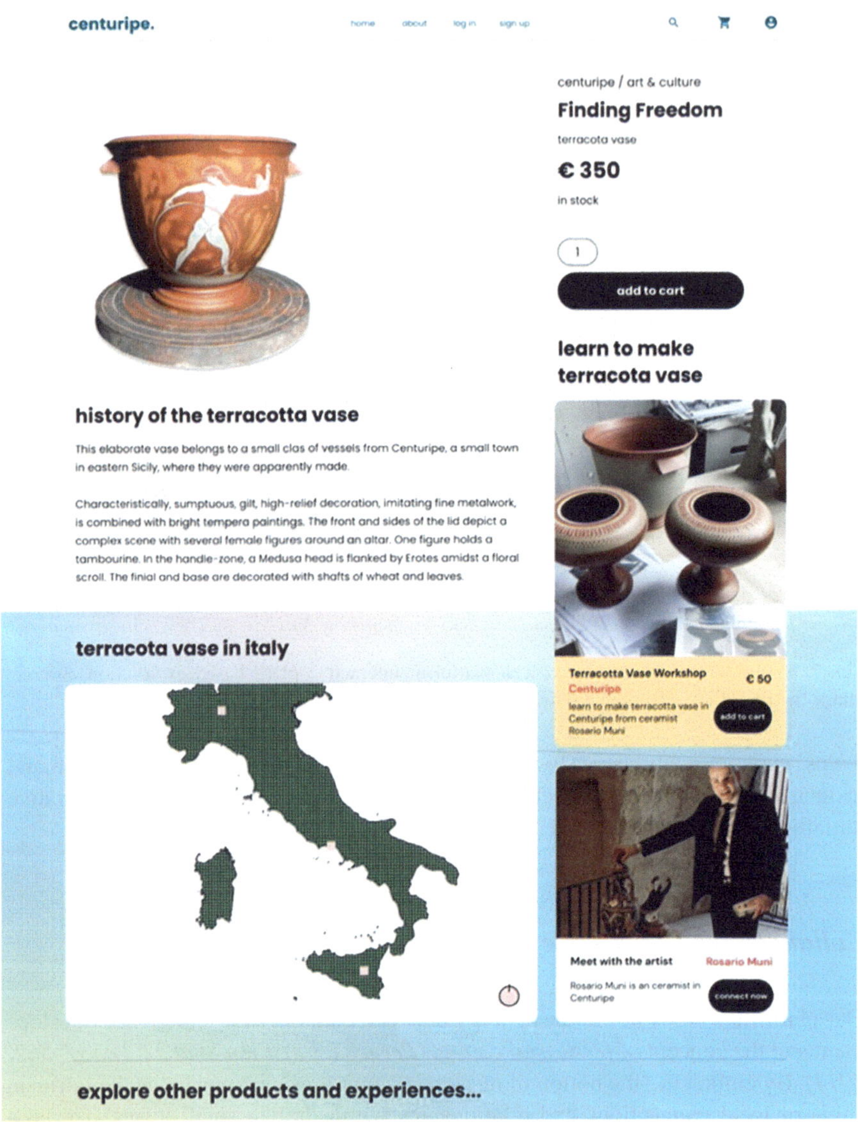

Fig. 8 Simulation of the proposed web platform: direct connection with small business owners. Source: image by the authors

Discussion

Our platform addresses a gap in the current market for place-based travel guides and application in Italy. While there are several location-based web and mobile applications to support businesses and tourism, there are limited web applications that focus on enabling local businesses to promote Italian small towns. Additionally, our application encourages connections among small businesses.

While we have outlined the initial stages of our digital platform's development for Italian SMEs, a journey of comprehensive research and refinement lies ahead. Our upcoming steps involve gaining a profound understanding of the needs of potential customers who seek authentic and offbeat experiences in Italian small towns. Firstly, we need to understand the needs of the potential customers that might visit Italian small towns. This user segment will likely include local and international tourists who seek "off the beaten path" experiences. We want to conduct in-depth interviews with this consumer segment of our target users to get specific data on their experiences including:

- Their typical behavior when selecting and planning travel to a destination.
- Current digital platforms and tools that support travel decision-making.
- Challenges and pain-points in their travel planning journey.
- Critical and unmet needs related travel decision-making and tourism.
- How small businesses at a destination play a role in their travel experiences.

Upon completing these user research, we will be able to establish product applications—comprised of specific metrics for our digital product that we will compare with major competitors. Some initial metrics that we can use to measure the impact of the intervention are the number of consumers visiting the platform, the number of small business owners using the platform, the number of connections made between small business owners and sales among SMEs.

With these insights as a foundation, we will establish precise product applications and metrics, enabling us to effectively gauge the impact of our intervention. We intend to rigorously test our platform assumptions through user testing, refining our design and features to ensure optimal user experiences. Additionally, recognizing the scarcity of open-source quantitative data to support our research questions, we plan to conduct our own comprehensive survey among Italian SMEs. This data collection initiative will delve into digital technology adoption, challenges in digitization, promotional strategies, customer dynamics, and economic trends, enriching our understanding and guiding the core features of our platform. In addition to demographic data about SMEs, we would also seek to answer:

- Digital technology adoption and impact among SMEs.
- Challenges with digitization for SMEs.
- Current behaviors to promote their business.
- Customer base.
- Changes in revenue and economic conditions over time.

These data would complement the qualitative data we collected among SME owners. Together, these data will provide a more comprehensive picture of the economic and digital development of SMEs and inform key features of our platform. Through collaborative efforts, innovative design, and strategic partnerships, we aspire to uplift local economies, empower SMEs, and provide a transformative platform that redefines the future of Italian tourism and economic development.

Conclusion

This project seeks to address the significant economic challenges posed by Italy's aging population and shrinking villages, particularly in the Lombardy and Sicily regions. By understanding the intricate relationship between digital activity, internet connectivity, and small business development, we aim to create a transformative impact on the local economies of Sondrio and Enna province. Through engagement with small business owners, we have identified a critical gap in digital adoption and recognized the importance of the *territorial capital, (Perucca, Eur Plan Stud 22(3):537–562, 2014)*, highlighting the interconnectedness of SMEs within local ecosystem. Consequently, we proposes a holistic web platform that showcases the unique stories and characteristics of Italian SMES, thereby enabling expansion and growth.

Our digital intervention is strategically aligned with various ongoing initiatives aimed at fostering economic revitalization throughout Italy. Drawing inspiration from these established endeavors, we are meticulously crafting our implementation strategy to ensure the success of our product. At the forefront of our approach, we have recognized the substantial investments made by the Italian government to uplift small towns and communities. Leveraging this momentum, we intend to seamlessly integrate our product with Google Maps, thereby enhancing accessibility and visibility for these lesser-known locales. Furthermore, our strategy encompasses the upcoming Winter Olympics, presenting a unique opportunity to showcase the cultural richness of Italy's hidden gems, such as captivating Valtellina district.

As we forge ahead, our overarching ambition is to catalyze the globalization of local economies, casting a spotlight on the charming tapestry of small Italian towns while propelling overall economic advancement. Notably, we acknowledge the pervasive digital divide that persists between SMEs in both the northern and southern regions of Italy. Even the most robustly performing Italian regions face a considerable gap when compared to their counterparts across Europe.

In light of these challenges, our implementation strategy is rooted in a commitment to bridge this divide. By leveraging cutting-edge technology, tailored support, and targeted outreach, we aim to empower SMEs across the nation, fostering inclusive growth and harmonizing economic development between different regions. Our journey is one of collaboration, innovation, and determination, as we navigate towards a future where Italy's economic landscape is transformed, and its diverse local communities thrive in the digital age.

Acknowledgements We do thank Isaac Castellani (Cà Rossa B&B, Montagna in Valtellina), Davide Fasolini (Dirupi, Montagna in Valtellina), Giuseppe Grasso (Allevabio, Vizzini), Salvatore Grifo (saffron farmer, Centuripe), Mattia Gurini (Cucina Yugen, Sondrio), Armando Lanzetti (Wine Hotel Retici Balzi, Poggiridenti), Niccolò Lenoci and Federico Romeri (food-tech startup Pascol, Sondrio), Rosario Muni (artist and ceramist, Centuripe), Silvia Scieghi and Stefano Canzi (organic and biodynamic farm Alpes Agia, Albosaggia), Savio Stancanelli (archeological ceramist, Centuripe). We extend our heartfelt gratitude to the dedicated individuals, SMEs, professors and faculty member from MIT Department of Urban Studies and Planning and Politecnico di Milano who have contributed to the success of our work.

References

Balocco, R., Cavallo, A., Ghezzi, A., & Berbegal-Mirabentet, J. (2019). Lean business models change process in digital entrepreneurship. *Business Process Management Journal, 25*(7), 1520–1542. https://doi.org/10.1108/BPMJ-07-2018-0194

ESPON. (2017). *ESPON policy brief on shrinking rural regions*. ESPON.

European Commission. (2022). Digital economy and society index 2021.

European Investment Bank. (2021). *The digitalization of small and medium-sized enterprises in Italy: Models for financing digital projects*. Publications Office. Retrieved November 8, 2024, from https://data.europa.eu/doi/10.2867/334596

Perucca, G. (2014). The role of territorial capital in local economic growth: Evidence from Italy. *European Planning Studies, 22*(3), 537–562. https://doi.org/10.1080/09654313.2013.771626

Tietjen, A., & Jørgensen, G. (2016). Translating a wicked problem: A strategic planning approach to rural shrinkage in Denmark. *Landscape and Urban Planning, 154*, 29–43. https://doi.org/10.1016/j.landurbplan.2016.01.009

Revitalize Centuripe: A Data Toolkit to Help Identify Buildings Across the Occupancy Spectrum

Ipshita Karmakar

Introduction

Italy is facing a rising problem of shrinking remote villages, declining, and aging population, and migration. The consequences of these multi-faceted factors have led to an increasing rate of building vacancy in Italy. Following in the footsteps of Mediterranean countries like Spain and Malta, it is estimated that building vacancy Italy would reach 20% overall by 2070, and has already reached a rate of 14.6% in the region of Sicily (Gentili & Hoekstra, 2019). National, local and territorial governments have yet to establish a building vacancy rate for across the country Istat (2022). The nature of this building vacancy, it's complexities and profiles have not been studied in great detail, even though this is rising source of alarm for local authorities in Italy who are dealing with the detrimental effects of this national problem at scale (Moroni et al., 2020).

Profiling Building Vacancy

Over the course of the field visits in 2022–2023 across 10-day intervals across Italy, one could observe several profiles of building vacancy across the country, that reflected each territory's socio-economic status and cultural specificities (Fig. 1). Additionally, in depth interviews with stakeholders across 7 towns in Sicily and Valtellina were taken with project partners from Politecnico di Milano, Abruzzo and Sicilian experts. In the Valtellina region, we identified a propensity of building vacancy due to the presence of several second/vacation homes. These well-kept

I. Karmakar (✉)
MIT Masters of City Planning 2023, Cambridge, MA, USA

G. Fossa et al. (eds.), *Small-Town Renaissance*, Research for Development,
https://doi.org/10.1007/978-3-031-89786-3_4

Fig. 1 Profiles of building vacancy across Italy. Clockwise from left: Sondrio, Valtellina; Vizzini, Sicily; Teglio, Valtellina; Favara, Sicily. Source: photos by author

buildings were used periodically over summer or skiing season but were otherwise well maintained and secure. In Sicily, we saw several abandoned towns, and buildings in various states of degradation, with several at the point of collapse. In the Abruzzo region, we observed retrofitted homes that were used seasonally, marked by boarded half doors to keep away animals in the meantime.

As per the Italian Revenue Agency (Agenzia delle Entrate), abandoned buildings are a subclass of unused buildings which, in turn, are a subclass of those that are at the owner's disposal (but not used continuously), which includes second homes. This definition therefore subsumes the magnitude of building decline and encompassed not just abandoned but also unused buildings, which may be used once in a while for short durations.

No known method existed in Italy at that time to document these various types of building vacancy and identify those which needed immediate attention.

The Problem of Unidentified Building Vacancy

Unidentified building vacancy in shrinking villages may risk several impending hazards. For instance, unoccupied and degraded buildings within heritage cores and particularly those that are a part of attached row houses face risk of collapse due to lack of maintenance and consequently tend to exert forces or degrade adjoining buildings.

Not all losses are tangible and visible. Any building that remains unoccupied in good condition is real estate that lies fallow. In a market with rising property values, one would assume that this is an opportunity cost, especially for real estate brokers and evaluators. No known national register exists in Italy for the real estate industry to evaluate abandoned and available properties (Henderson, 2020).

Each abandoned building also contributes to the overall degradation of the economic status of the village or town that it exists within. The overall standard of living with the town is compromised, and the taxation that contributes to functioning public services and amenities reduce as well.

Small villages within rural Italy also lack adequate staff capacity to conduct door to door surveys and quantify building abandonment. According to the mayor of Centuripe, there are no known records of property being abandoned, water tax or meters being paid, or property tax being paid in Centuripe. There are only about 12 full time staff members in Centuripe city council, and quantifying building abandonment, while a high priority problem, unfortunately doesn't get the number of man hours it requires.

Italy's Rural Settlements Are Shrinking

The reason for this rising abandonment in rural Italy seems to arise from several overlapping pressures—there is rising emigration from Italy to other European countries, and within the country itself, there is a rising internal migration to larger cities (Moroni et al., 2020). Within rural Italy, a rapidly increasing aging population is bequeathing property to a generation that does not exist, or to too many competing property eligible heirs who prefer to not engage in property battles. The cost of inheriting and maintaining large palazzos for nuclear families is also a hassle that not many want to engage in. It is easier therefore, to let a building go into ruination than to engage meaningfully in property transfers, sales or demolition.

Opportunities for Rethinking Abandonment

In order to tackle this rising issue of building abandonment and the phenomenon of 'ghost towns' in rural Italy, the Italian government is actively pursuing European Union grants and leveraging funding from the National Recovery and Resilience Plan (2021), that pumped millions of euros into the Italian economy, particularly for small town revitalization post covid.

According to a 2022 Forbes report, Italy is channeling almost one billion euros into revitalizing shrinking villages by incentivizing the stay of digital nomads with increased amenities and infrastructure upgrades (Hughes, 2021). Additionally, villages with a population under 10,000 residents are eligible for '*I borghi più belli d'Italia*' designations, that classify them as heritage beautiful 'little' Italian villages which attract tourism.

This abandoned property is ripe for real estate investment by private parties, to encourage transactions within these towns. However, in order to do that, rural Italian villages first need to invest into upgrading infrastructure and amenities, and creating an environment that is conducive to higher demand and better regulation. There have been offers to sell houses within rural Italy for as low as $1, and there have been little to no takers so far. However, in cities such as Milan, rental rates shoot upwards of $500 for decent property. Italy's rural villages have the possibility of attracting these city residents with cheaper property rates and better transportation, if a conducive regulatory and governance environment is engineered.

Most villagers within these shrinking villages are not only acutely aware of the problems that their villages are facing but are also one of the best agents of change within their villages. Interviews in Favara, Sicily and Catania, Sicily with local, entrepreneurial organizations such as South Working (an NGO based in Sicily) reveal an innate knowledge by local Italians of not just disinvested properties, but also underinvested rejuvenating activities that can help revitalize their own villages.

The lack of information regarding these properties can also be remedied with better governance and data management systems. In today's day and age, there is also a proliferation of publicly available data such as Google Street Views, Open Street Maps data, Flickr images or other social media that provides us an overview of the nature of a rural settlement in Italy. These datasets may be biased or inaccurate in some instances, but the sheer quantity of these data sets help override some of the anomalies or inaccuracies one may find within the data and help with exploratory analysis.

The availability of this data also uncovers the possibility of using emerging technology such as models of machine learning and visual recognition to identify patterns from this data and predict, to a certain extent, certain algorithmic combinations that point towards abandonment, which in a way help tide over the lingering issues of lack of staff and administrative capacity in small municipalities.

Research Questions

The research therefore asks two interrelated questions pertaining to both data collection and social governance in shrinking cities of rural Italy. First, could image recognition techniques be used to identify profiles of abandonment and occupation of buildings in shrinking rural towns of Italy, particularly principal component analysis for visual indicators? Second, could civic engagement be encouraged for crowdsourcing data on vacancy, solicit ideas and sustain long term push for policy change?

Data and Methods

The lack of publicly available data on building abandonment at national, territorial, and local levels, including the non-availability of property tax data led this research to create its own methods of data collection and data evaluation, while also relying on public data sets such as google street views. This was achieved by tabulating a crowdsourced data using mobile forms and an educational web platform and processed using principal component analysis. This dataset was later used to train a machine learning model that could identify building abandonment from google street view images and other public datasets. A pilot study was conducted in Pettorano sul Gizio in Abruzzo, a village of around 600 people, and on 50 lots in Centuripe, Sicily.

In order to streamline the process of data collection and data processing, the process of data collection, and the method of machine learning had to be refined.

Designing the Process

This methodology would work best for inner city towns in Italy which would most benefit from policy interventions, have high potential for impact, and lack the administrative capacity to classify building occupation. These mixed methods approach also relied on establishing prior relationships to the city council and community to build trust before introducing the minimally invasive process of data collection within the town/village/city. Combining datasets collected with this methodology with other datasets from other oral accounts of building vacancy and other crowdsourced data is also highly recommended. Further, this research relied on a two-way information transfer between the community and researchers, which included the researchers providing a constantly updating, publicly available dataset to the community, along with a portal for the community to raise suggestions and objections to the data collected, and an opportunity for the community to generate ideas regarding these abandoned properties for their alternative use.

Designing the Method

Visual recognition methods for processing this algorithm using machine learning requires a standardization of imagery i.e., handheld images, google street view images, and 360 camera captures have all different resolutions and angles, and this creates noise which makes it difficult to run a visual recognition algorithm. Therefore, in order to first train the visual recognition model, there needed to be a standardization of images for at least 100 images before the algorithm can be tested on datasets such as google street view images (Figs. 2 and 3).

Additionally, relying only on a machine learning model without supplementing it with local knowledge can lead to bias. Digital accessibility in most rural Italian villages is also less than 70% in certain cases, and purely digital data sources may either not be available at all, or intermittently available.

As a result of all of these limitations and data gaps, the research creates its own initial datasets based on its own criteria for visual recognition, based on selected visual indicators.

Based on initial observations and walking tours of Centuripe, Sicily and Pettorano sul Gizio, 18 visual indicators were identified to relay the level of abandonment on a spectrum of complete occupation to complete ruination.

Fig. 2 Google street views from Centuripe in Sicily, and a handheld image from the same area. Source: Google Earth; photo by the author

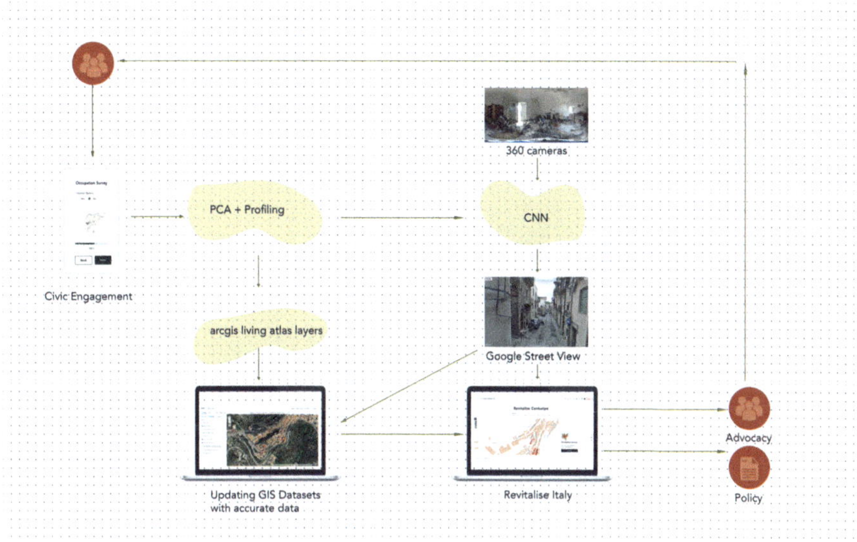

Fig. 3 Data collection method and pipeline for the research project including software requirements. Source: image by the author

- Collapsed roof.
- Dead plants.
- Tension ties.
- Spalling.
- Broken doors.
- Plant infiltration.
- For sale signs.
- Half board.
- Padlock.
- Restored fixtures.
- New paint/plaster.
- Kept plants.
- AC/satellite.
- Names on buzzer.
- Exposed pipes.
- Laundry.
- Trashcan.
- Sounds of life.

Each of these isolated visual indicators is indicative of a state of housing degradation. Each of these was also selected with a certain preconceived assumptions i.e. kept plants relates to a building in occupation and collapsed roof relates to a building in ruins. For the purposes of creating the first data set, the researchers created a survey based on a simple mobile form that relied on manual inputs of visually assessed data by individual data collectors. The presence of each indicator was coded as 'yes' or 'no' on the mobile form, therefore leading to a dataset of 18 binary variables for each building.

Pilot Study: Pettorano sul Gizio

The pilot study was conducted in Pettorano sul Gizio, a town of roughly 800 families in the Abruzzo region of Italy. This pilot study was conducted along with local research group Liminal, and fellow students Adriana Giorgis and Lauren Gideonse, M.Arch'24 at MIT School of Architecture and Planning, with the support of the mayor of Pettorano sul Gizio and community members. The three researchers chose a sample set of around 230 houses to visually assess using the above mentioned mobile form of 18 visual indicators, and took around three days to assess three transects around the town (Figs. 4 and 5).

This study resulted in a first pilot analysis of this dataset, using a simple filtering function to distill the groups of visual indicators which best described the spectrum of building occupancy i.e., Buildings predicted to be in complete occupation, seasonally used buildings, recently abandoned buildings, and buildings in complete ruins.

This study was further refined when it was tested on 4 towns in Lazio, near Rome, by students at the University of Miami, who adopted the same mobile form survey and tested them on a broad spectrum of settlement sizes, densities and typologies across different towns, with only one prior remote training session by the survey developers.

During this exercise, researchers asked the third-party enumerators (students) questions regarding ease of use, adaptability of this survey in different contexts, and willingness of locals to engage in the survey, in order to determine the feasibility of a crowdsourced survey built out of the contributions of locals themselves.

Site Context: Centuripe, Sicily

Centuripe is a quaint village in Sicily, at an elevation of 2402 feet on a ridge between Simeto and Dittaino rivers, northwest of Catania (Fig. 6). This village of around 5000 residents has around 3800 buildings, and an institutional presence of high

Fig. 4 Mobile data forms and qualitative walking tours of Pettorano sul Gizio. Source: images by Ipshita Karmakar, Adriana Giorgis, and Lauren Gideonse

Fig. 5 Engagement with University of Miami students for fieldwork in Lazio using the same survey forms. Source: screenshot by Liminal Research group

schools, museums, and a vocational training center. Salvatore La Spina, mayor of Centuripe at the time, estimates around 191 abandoned houses in Centuripe, however, no official count of abandonment exists within the city council, and this count is assumed to be grossly underestimated (Fig. 7).

Since the initial count of the population of Centuripe is well below 10,000, Centuripe is a good candidate for NRRP funds for declining villages that are eligible for revitalization grants. In addition, Centuripe is a fifth century Hellenized commune with Greek and Roman influences within its architecture, and can additionally target cultural heritage grants to revive and maintain its archaeological and historical glory. In June 2022, researchers from MIT DUSP conducted a transect study in a neighborhood to understand the nature of building vacancy and conducted initial interviews with stakeholders on site. In addition, photographic evidence of abandoned properties and 360 camera footage of interiors and exteriors of a few selected sites were obtained.

Since initial conditions proved favorable for Centuripe as a study area, the research chose to continue its investigation into the commune beyond initial field studies and across 4 months in 2022 to test the above-mentioned survey using crowdsourced data.

Data Collection

Data collection support was requested from the mayor's office in Centuripe, keeping in mind limitations with staff capacity. Francesco Puglisi, an employee of the Regional Archaelogical Museum at Centuripe and at the city council volunteered 4

Fig. 6 View of Centuripe. Source: photo by Brent Ryan

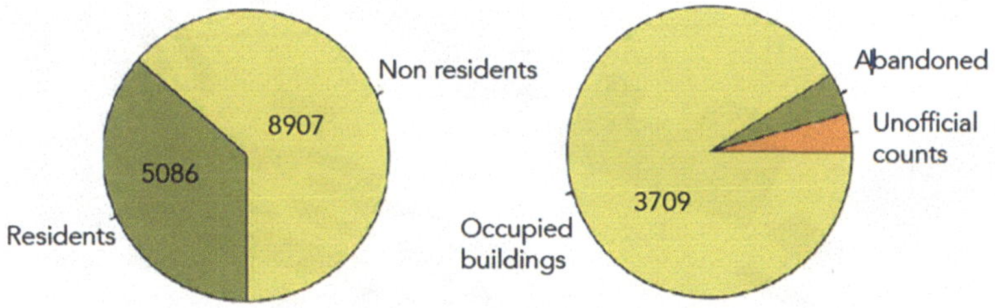

Fig. 7 Data related to population and buildings in Centuripe. Source: image by the author; data: City of Centuripe

h of time on field to test the building occupancy survey. With only half an hour's remote training, Francesco was able to survey 50 lots across a transect in a predetermined neighborhood. In addition, he was able to contribute his own knowledge as a local and ask for feedback and inputs from other locals to verify and ground truth the survey and help add a few additional variables to the survey (Fig. 8).

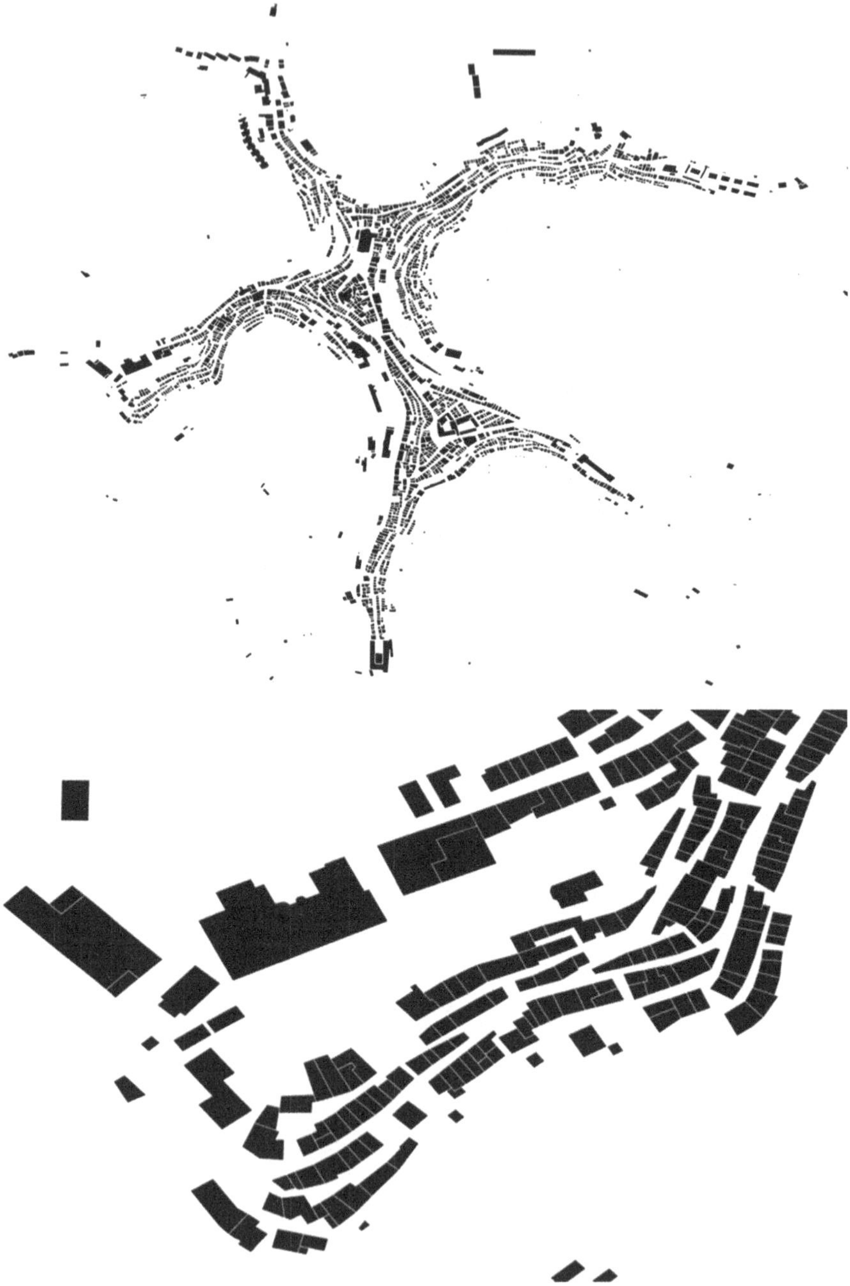

Fig. 8 Surveyed areas of centuripe. Source: image by the author; data: Open Street Maps

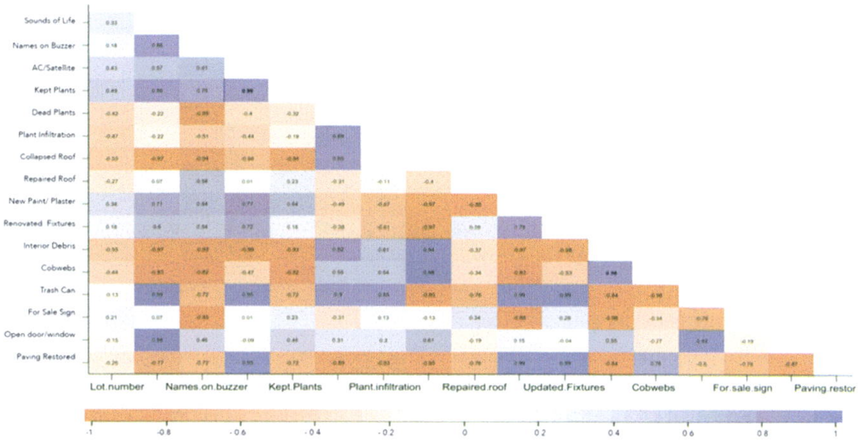

Fig. 9 Tetrachoric correlations between survey items. Source: image by the author

Analysis

We used principal component analysis to distill the results of our survey into broad categories within correlated components of our 17 variables. We used a tetrochoric correlation matrix to identify collinearity between our indicators. Based on our correlation matrix, we can observe that indicators such as collapsed roof and dead plants have an 85% positive correlation, and indicators such as sounds of life and interior debris have a 97% negative correlation. This test is useful in determining clusters of indicators which may all point to a particular trait of building occupancy (Fig. 9).

We ran a principal component analysis[1] to identify the highest eigenvector values of 16 components (Fig. 10).

The presence of cobwebs, collapsed roofs and interior debris have the highest weights and the presence of new paints, sounds of life, and kept plants have lowest weights in the first principal component.

Since we knew that the data set was biased towards unoccupied buildings, we can say that the first principal component can be identified as the weighted sum of indicators that identifies ***building abandonment***.

We proceed to tie the column containing principal component 1 from the tetrachoric correlation matrix back into our original data frame of 50 lots. Using this method, we can rank the lot numbers on a scale of principal component 1 (ranging towards an indication of abandonment).

[1] The author adopted the following algorithm: *C1= -0.27zsoundsoflife-0.28znamesonbuzzer-0.31zac.satellite-0.28zkeptplants+0.28zdeadplants+0.27zplantinfiltration+0.32zcollapsedroof-0.08zrepairedroof -0.29znewpaint-0.31zupdatedfixtures +0.32zinteriordebris+0.29zcobwebs -0.12ztrashcans-0.03zforsalesigns+0.03zopendoor -0.14zpavingrestored.*

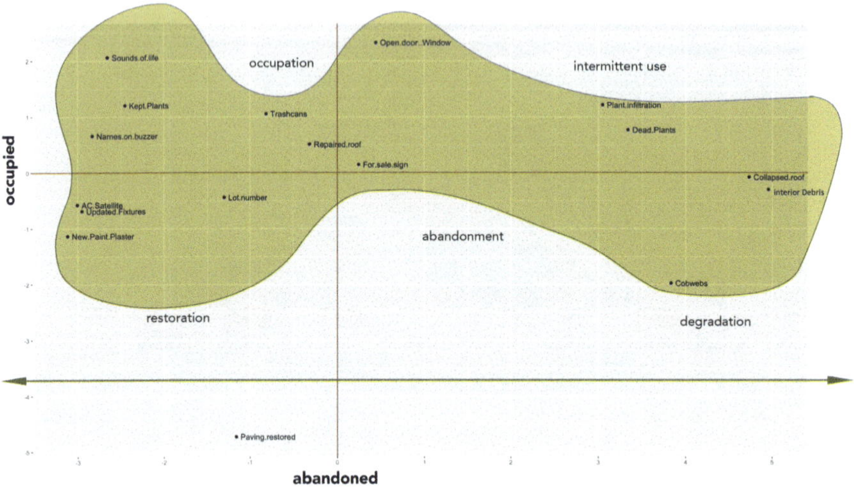

Fig. 10 Distribution of first principal component (X axis) and second principal component (Y axis). Source: image by the author

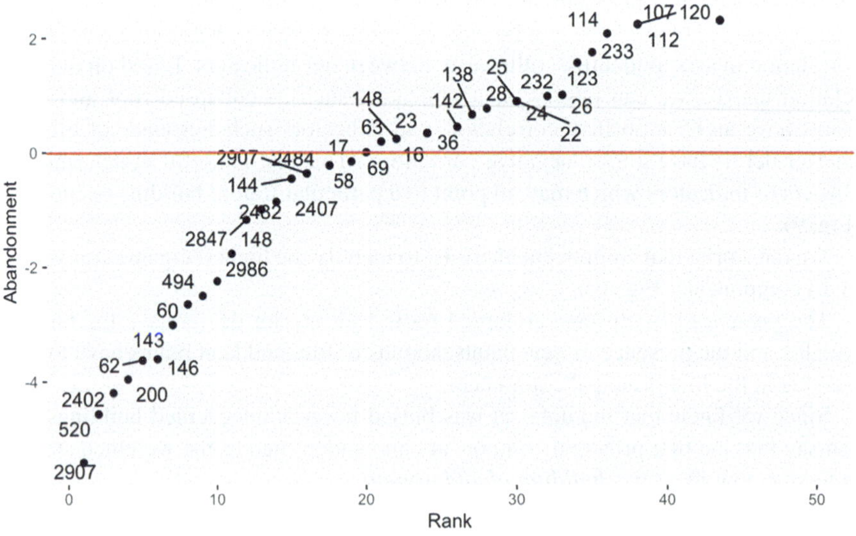

Fig. 11 Ranking of buildings across the first principal component i.e. abandonment. Source: image by the author

From the above graph (Fig. 11), we can determine that lot 120 is the 'most unoccupied' on the spectrum of building abandonment and lot no. 2907 is the most 'occupied'. By ground truthing our data on the site using qualitative surveys and corroborating it with local authorities, we can confirm that lot 120 is unoccupied and abandoned. Lot no. 2907 is a local school that is currently in use. This confirms our survey validity and ensures that all lots with a positive value for principal component one should be in some level of physical degradation and therefore should require human intervention and a further validity check (Fig. 12).

Since the survey required a geolocation and a geostamped image to be uploaded corresponding to each lot, we were able to determine the location of each abandoned building corresponding to Principal Component 1 on a map. We were able to take the point coordinates of each identified lot and conduct a spatial join with building polygons in our cadastral map (Fig. 13).

Fig. 12 Examples of images collected as a result of mobile form surveys, which can be used to train a convolutional neural network model to train image recognition algorithms. Source: photos by the author

Fig. 13 Abandoned lots as identified by the building occupancy survey in Centuripe. Source: image by the author

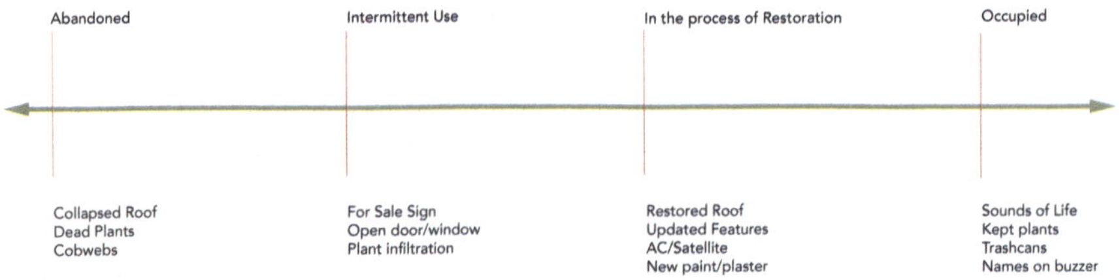

Fig. 14 Spectrum of building occupation based on clustering of visual indicators. Source: image by the author

Further, based on the clustering of variables in the principal component analysis and the tetrachoric correlation, we were able to create a spectrum of indictors for building occupancy (Fig. 14).

We used the following indicators and help create three more profiles of building occupancy—intermittent use, in the process of restoration and occupied. We used Principal Component 16 to identify the occupied buildings, and a separate if/then analysis[2] to identify intermittently used and partially restored buildings.

The following are the geolocated lots that were identified to be intermittently used, restored or completely occupied (Figs. 15, 16, and 17).

[2]The author adopted the following algorithm:

Intermittently used = if (and (for.sale.sign==1),(open.door.window==1),(plant. infiltration==1),1,0).

Restoration = if (and (restored.roof==1),(updated.features==1),(ac.satellite ==1),(new.paint. plaster ==1),1,0).

Fig. 15 Intermittently used buildings as identified by the building occupancy survey of the 50 lots sampled in Centuripe. Source: image by the author

Fig. 16 Restored buildings as identified by the building occupancy survey of the 50 lots sampled in Centuripe. Source: image by the author

Fig. 17 Occupied buildings as identified by the building occupancy survey of the 50 lots sampled in Centuripe. Source: image by the author

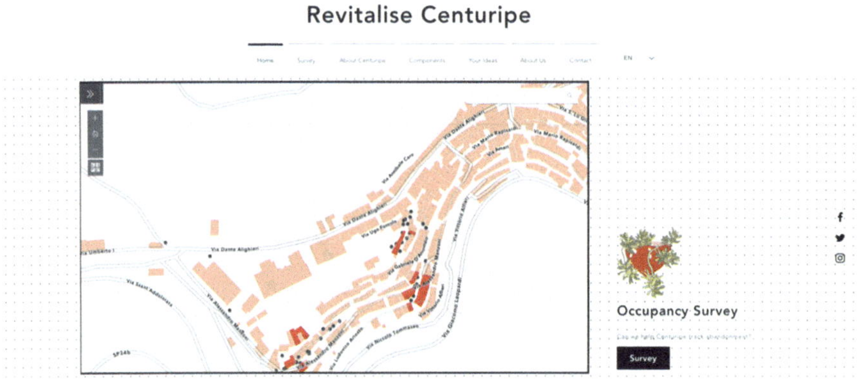

Fig. 18 Revitalize Centuripe is a web platform created by the author that hosts the building occupancy survey for Centuripe and its results. Source: image by the author

Possible Uses of this Data

Two-way information transfers between researchers and community members could be enabled by creating a web platform where algorithmic information generated could be visualized, updated, debated upon and verified as needed. Therefore, a city specific website www.revitalisecenturipe.com was created with a copy of the survey, a portal to verify whether the information that was generated using the algorithm was true, and a space to generate ideas regarding abandoned properties by several stakeholders with competing claims (Fig. 18).

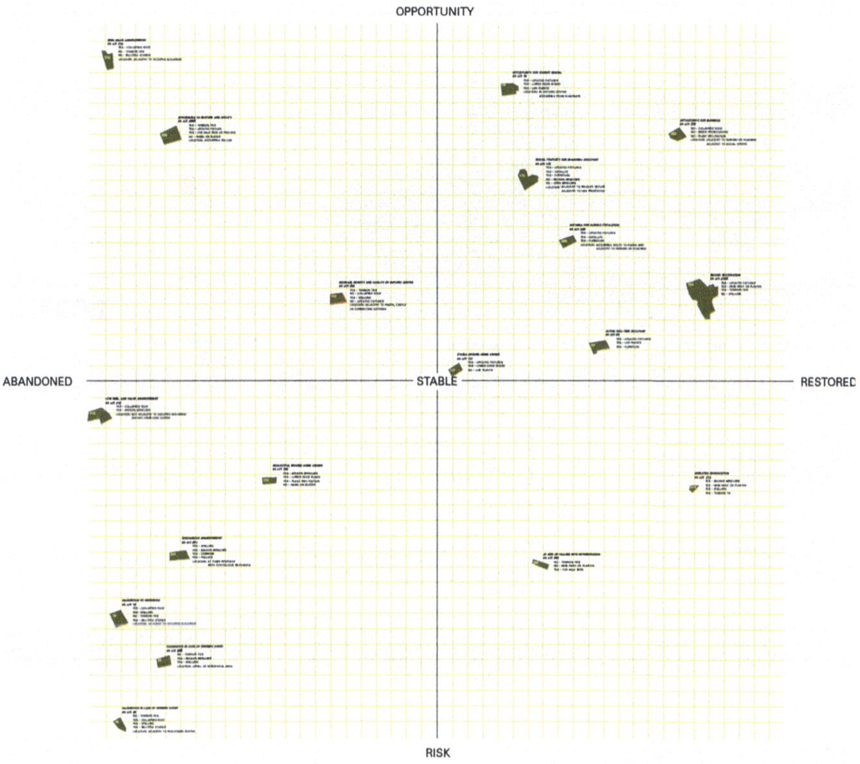

Fig. 19 Risk matrix of risk and opportunity for identified abandoned buildings in Pettorano sul Gizio. Source: image by the author

This information on building vacancy can be used to create a method to identify at risk buildings and direct care and resources towards those that are close to the risk of collapse. In order to identify the most at risk buildings, a matrix of risk vs opportunity was created to identify buildings that should be prioritized for restoration (Fig. 19).

Propositions and Way Forward

While the results of this algorithmic, big data generated and small data verified information for Centuripe was successful to an extent, the data had limitations in accuracy and the costs of the exercise far outstripped the benefits since only 50 houses were surveyed after many negotiations with city council officials. This research suggests several future directions that this project could undertake to make this study more academically sound and increase the likelihood of a crowd generated survey.

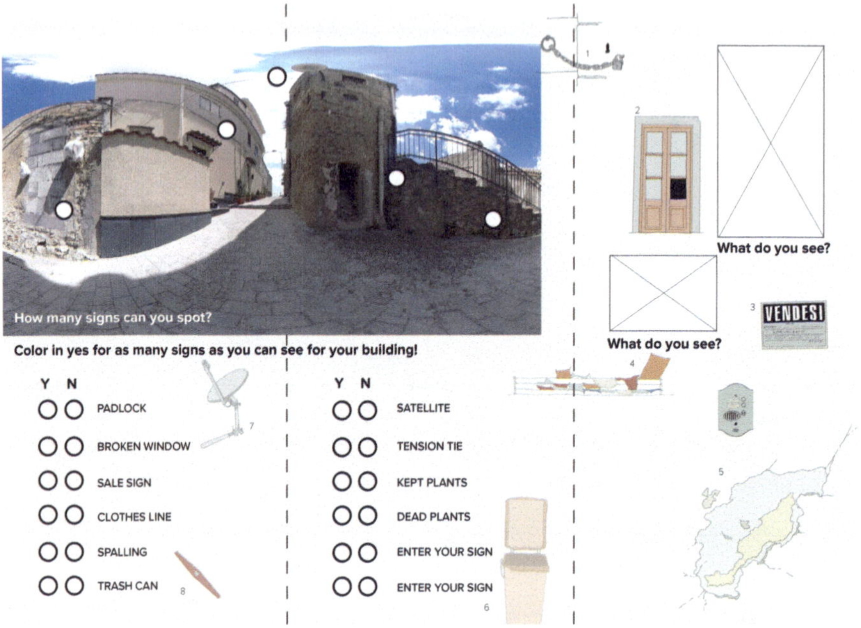

Fig. 20 Paper pamphlet version of the building occupancy survey for Centuripe. Source: image by the author

Analog Applications

Even though the Italian government had set a goal of providing ultra-fast broadband to 85% of the population, data reach and data accessibility in remote rural areas in Italy has been challenging (Frąckiewicz, 2023). In order to overcome this challenge, the research develops an analog method using paper pamphlets designed as an engaging game, that can provide the same information, without relying on an internet connection (Fig. 20).

Furthering Community Engagement

To incentivize further engagement with this building occupancy survey, it is important to invest in education programs on understand the phenomenon of shrinking villages in Italy, and the impacts of this shrinking in one's own village. In the case of Centuripe, the Municipality could engage the Regional Archaeological Superintendence, museums (Fig. 21), schools and other universities to participate in this survey funded by cultural and educational research grants.

Fig. 21 Regional archeological museum, Centuripe. Source: Wikicommons

Conclusions

This research is conceptualized as a granular village wide study that can be aggregated to create a municipality wide roster of building occupancy, that can further be aggregated to create a nationwide roster of building vacancy. However, the benefits of this study should be used with discretion, since this data set of property generated non-invasively can create problems of misidentification (solved in part by those who have access to the web platform), provide discriminatory data that could be used to persecute low income families that cannot afford upkeep of their houses, and be used by real estate agents to buy out ancestral properties. Keeping these limitations in mind, this research provides a method to those municipalities who are willing to regulate, fact check and benevolently use building occupancy data to help revitalize their villages economically and culturally.

Acknowledgements The author would like to acknowledge the support of MIT's Department of Urban Studies and Planning, MIT Department of Architecture and Liminal for arranging field visits to Italy in 2022 and for Politecnico di Milano professors and faculty members for their expert support and guidance. Further, the author would like to thank Francesco Puglisi (Regional Archeological Museum, Centuripe), and MIT M. Arch students Adriana Giorgis and Lauren Gideonse, who were the co-creators of the first building occupancy survey tested in Pettorano sul Gizio, Abruzzo.

References

Frąckiewicz, M. (2023). *Internet access in Italy*. TS2 SPACE. Retrieved May 19, 2024, from https://ts2.space/en/internet-access-in-italy/

Gentili, M., & Hoekstra, J. (2019). Houses without people and people without houses: A cultural and institutional exploration of an Italian paradox. *Housing Studies, 34*(3), 425–447. https://doi.org/10.1080/02673037.2018.1447093

Henderson A. (2020). *Why I'm not buying cheap real estate in Italy or Spain*. Nomad capitalist. Retrieved November 8, 2024, from https://nomadcapitalist.com/finance/investing/dont-buy-cheap-real-estate-in-italy-spain/

Hughes R. A. (2021). *Italy's €1 billion project to lure digital nomads to remote villages*. Forbes. Retrieved November 8, 2024, from https://www.forbes.com/sites/rebeccahughes/2021/04/26/italys-1-billion-project-to-lure-digital-nomads-to-remote-villages/

Istat (2022). Retrieved June 9, 2025, from https://esploradati.istat.it/databrowser/#/en/dw/dashboards

Moroni, S., De Franco, A., & Bellè, B. M. (2020). Unused private and public buildings: Re-discussing merely empty and truly abandoned situations, with particular reference to the case of Italy and the city of Milan. *Journal of Urban Affairs, 42*(8), 1299–1320. https://doi.org/10.1080/07352166.2020.1792310

National Recovery and Resilience Plan. (2021). Retrieved June 9, 2025, from https://www.italiadomani.gov.it:443/content/sogei-ng/it/en/strumenti/documenti/archivio-documenti/national-recovery-and-reslieince-plan.html

A Sport and Heritage Corridor for the Valtellina Olympics

Giovanna Fossa, Felix Günther, and Annalisa Rollandi

Vision

The 2026 Winter Olympics raised the long-standing problem of connecting Milan to Valtellina, where ski competitions will be held (Bormio and Livigno), as a key priority: it is anticipated that the event will lead to a significant increase on the current tourist flow (Fondazione Milano Cortina 2026, 2024).

The present proposal extends beyond the mere necessity of mobility during the Olympic event between Milan and the surrounding mountains. It aims to reinterpret mobility in Valtellina within a framework of multimodal sustainability and a non-sectorial, integrated strategy for settlement regeneration at a multiscalar level. This proposal is designed to systematically integrate the sports-tourist offer with the historical and cultural identity that distinguishes the valley, thereby enhancing its uniqueness and excellence.

The Olympics provide an opportunity to experiment with a future spatial layout of the valley and to allow visitors and the population to enjoy a novel territorial experience. Valtellina's linear configuration, situated at the heart of the Alps, lends itself to the concept of a transalpine corridor to enhance the region's potential in terms of sustainable and experiential sports activities. This is a strategic vision of territorial regeneration, serving as a legacy of the Milan-Cortina 2026 Olympics.

G. Fossa (✉)
Department of Architecture, Built Environment and Construction Engineering, Politecnico di Milano, Milano, Italy
e-mail: giovanna.fossa@polimi.it

F. Günther · A. Rollandi
Department of Environment, Constructions and Design, University of Applied Sciences and Arts of Southern Switzerland, Mendrisio, Switzerland
e-mail: felix.guenther@supsi.ch; annalisa.rollandi@supsi.ch

G. Fossa et al. (eds.), *Small-Town Renaissance*, Research for Development,
https://doi.org/10.1007/978-3-031-89786-3_5

The sporting events that Valtellina hosts provide an opportunity to redefine its image within the context of a new, wide-ranging strategy. The combination of sport and heritage is an effective strategy for rebranding and territorial marketing: it responds to the post-pandemic demand for holistic physical and mental well-being by enhancing specific resources of Valtellina and its context of Alpine and pre-Alpine marginal territories. The usual metaphor of the corridor as the backbone of heavy mobility infrastructure is overlaid with the lighter infrastructure of culture, history, sports, and tourism, thereby facilitating a reinterpretation of an interconnected and integrated network of livability, recreation, and tourism.

In other words, a corridor of sustainable mobility integrated with innovative telecommunication services for sports and historical landscapes, is aimed at new trends of experience and adventure tourism, with the potential to involve smaller towns to reverse the trend of depopulation and abandonment.

The proposal is made technically possible by the continuous improvement of the digital coverage of the Valtellina and its side valleys (UFCOM, 2023), with the extension of the possibility of connection not only on the valley floor but also on the slopes and reliefs: satellite and fixed wireless accessibility technologies cover orographically complex areas not reachable by fiber optics,[1] guaranteeing quality and consistency of connection at professional levels. The technical conditions are thus in place for the development of digital platforms and applications that integrate multimodal mobility services with sports and tourism services. It is therefore possible to provide personalized services for an offer that is as rich as it is dispersed and complex to organize logistically and, above all, to know and imagine without innovative digital tools.

Displacement along, from, to or around the corridor, is conceived as a whole as an experience of psycho-physical well-being, so that it constitutes a destination in itself: a synesthetic experience nourished by landscapes and physical activity, culture and active recreation enjoyed by moving along historical paths or infrastructures. The fact that displacement becomes a destination in this perspective digitally extends and innovates the established model of the Bernina Express and the emerging focus on the "Via dei Terrazzamenti" (Terrace Route), both of which are UNESCO World Heritage Sites for the historicity of the infrastructural landscape (the Bernina Express for the century-old railway line that overcomes the most pronounced gradients with spirals supported by arches, the Terrace Route for the art of making dry stone walls of the infrastructural landscape).

In other words, the proposed sustainable mobility corridor is integrated with innovative services for sports and historic landscape offerings and oriented toward new sustainable active recreation and experiential tourism, with the potential to engage smaller towns and reverse the trend of depopulation and abandonment. The strength of this vision, in addition to the originality of thematic integration, is to open up to the macroscale, across the Alps, drawing upon the catchment area of the

[1] See chapter Boniotti, C., Pimpinella, A. "Heritage and Mobile Radio: A Data-Driven Approach to Address Cultural Heritage Valorization".

metropolitan regions of Milan and Zurich and positioning Valtellina as a hub between the destinations of excellence of Lake Como and Engadine/S. Moritz.

The corridor is developed by leveraging the Milan-Lecco-Tirano-Bernina railway line in conjunction with the water greenways along the Martesana canal and the Adda river, as well as the Swiss national bicycle and hiking network (Regione Lombardia, 2021; ARE, 2024). This multimodal corridor has the potential to systematize the sport environments and historical-cultural resources of Lake Como and Lake Lugano, Valtellina, and Engadine, while also fostering synergies between their respective product and landscape brands.

The concrete and feasible nature of this strategic vision is demonstrated, with a multiscale approach, identifying as a flagship project[2] the Smart Terrace Trail (i.e., the proposed transformation of the bike/pedestrian route "Via dei Terrazzamenti" into a backbone of experiential tourism and smart working).

From Milan, which boasts an Olympic arena and a plethora of major sports centers, the corridor traverses the landscape of rivers and canals of the dry plain, pre-Alpine hills and lakes, the Valtellina valley floor and cultivated slopes. It then continues through Valposchiavo and the Bernina Pass (the highest point which could name it as "Bernina Corridor"). It is recommended that the skiable domains of the upper Valtellina and Engadine be integrated. This would facilitate connectivity with the Swiss national railway network, which descends back over the watershed into the Grisons elevations, flanked by skiable domains, and ultimately reach the Zurich metropolitan area. This long-term planning vision is not limited to the winter season and takes into account not only tourist but also commuter flows in pre-Alpine and mountain contexts, including cross-border ones. It is a backbone of integrated services and an "experiential" corridor for sports-landscape enjoyment.

Materials and Design Criteria

The materials for this vision include sustainable mobility routes (railway lines, waterways, greenways, trails, and decommissioned historic routes), landscapes, built and cultural heritage, sports domains, sports facilities, health facilities, and tourism services.

The structural linearity inherent to the historic landscape of the Valtellina and Como lake incision presents an opportunity for the development of linear sports activities and longitudinal routes for active and passive recreation. Indeed, the topography of the territory has shaped over time a multitude of historic linear trails. Initially serving as communication, trade, and pilgrimage ways or roads/muletracks/trails supporting agro-silvopastoral activities, these routes have now been rediscovered or have the potential to be reused as sport-scenic itineraries.

[2] See chapter Fossa, G., Guenther, F., Paolini, D. "The Smart Terrace Trail as Strategic Regional Project".

These paths exhibit a parallel structure on the valley floor or lakefront and on the midslope. Examples of valley floor routes include the Valtellina Path (a bicycle path along the Adda) and the sequence of traditional promenades along the lake. The routes along the eastern side of Lake Como, between Abbadia Lariana and Colico, are collectively referred to as the "Wayfarer's Path" (Sentiero del Viandante) while the paths along the Rhaetian side of Valtellina are known as the "Terrace Route" (Via dei Terrazzamenti). These routes are optimal for walking or bicycling, although they can also be traversed on horseback, by canoe along the Adda River, and by sailing along the eastern shore of Lake Como. Additionally, there are parallel ridge routes on the lake and mountain watersheds of the Adda basin, including the "Sentiero Italia", which features ski mountaineering routes and high-altitude hiking/mountaineering traverses (Fig. 1).

Historic routes and infrastructures have the potential to serve as the foundation for heritage systems and sport-experiential trails, as well as for mental and spiritual regeneration. Some routes are Marian paths or already host mountain running or ski mountaineering competitions.[3]

The axis of sustainable mobility, intertwined with and supporting these linear sports routes, is identified in the Milan-Lecco-Tirano railway line (currently a minor line in the Lombardy network), in the lake navigation boats and in the historic roads (the old Spluga state road along the lake and the "wine road" along the Rhaetian terraces), freed from through traffic by the recent highways, which mainly pass through tunnels.

Access to the Valtellina is still largely by car, in an exclusive form for the middle and upper Valtellina above Tirano, where the Italian railway currently ends and the Swiss Bernina railway begins.

The regeneration of the junctions between different sustainable modes of mobility is considered strategic for the systematization of the already functioning corridor sections, drawing poles of ecosystem and people services, for inhabitants and tourists: localized services, added to the diffuse ones provided by apps, essential to guide tourists in personalized choices of fruition of the variety of heritage landscapes, sports activities and opportunities to meet with an authentic territory, from food and wine to crafts. The development of these polarities involves not only functional aspects of the mobility hubs, but also urban planning aspects. They will become centers of activity for both visitors and residents, with high-quality urban spaces that welcome visitors and allow them to enjoy a pleasant stay while changing from one mode of transport to another. A high density of offers and aesthetic quality will characterize these hubs.

The valorization of historical infrastructures as a cultural asset, exemplified by the Unesco Heritage designation of the Bernina railway route, is coupled with an understanding of the exchange of artistic and craft traditions within the Alpine region (Unesco Intangible Cultural Heritage designation of dry-stone walls). This sharing of architectural elements and techniques spans not only ancient times but

[3] See the annual Orobie Ski Mountaineering competition.

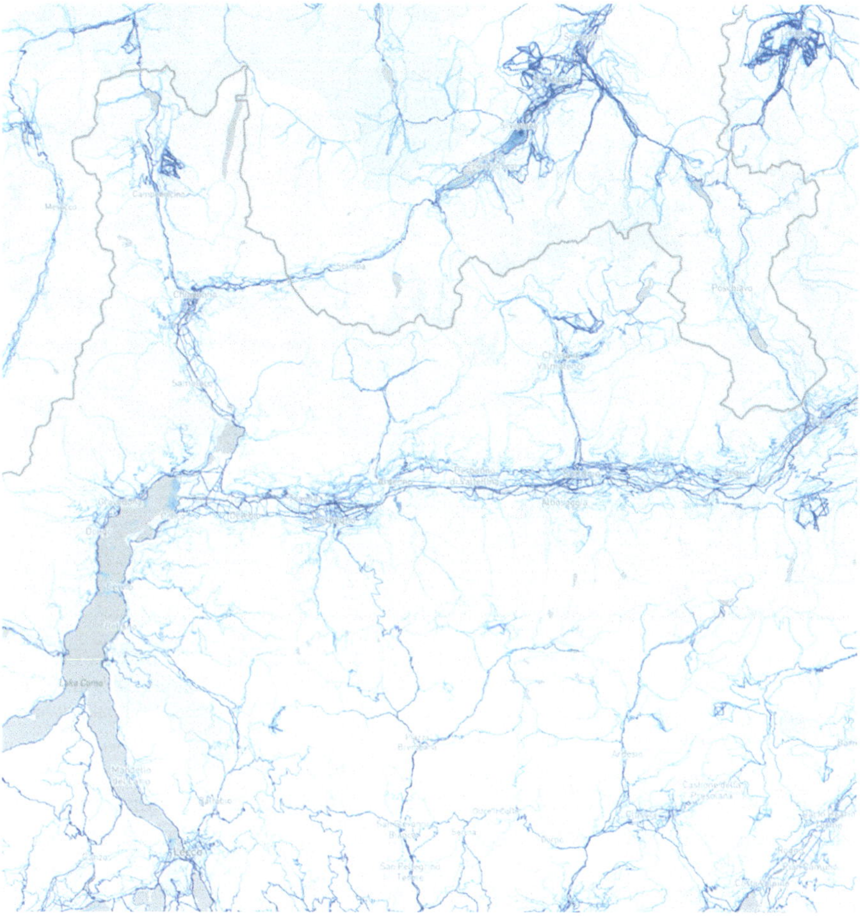

Fig. 1 Sports activities concentration along the Bernina Corridor. Source: Strava, Inc. (2024) *Strava Global Heatmap*. Retrieved from https://www.strava.com/heatmap

also recent ones, thus constituting a common Alpine cultural heritage (Fig. 2). The Bernina Corridor, for instance, has played a role as a spine of modern architecture: in his 1950s projects located along the Valposchiavo and Bassa Valtellina to Milan, arch. Luigi Caccia Dominioni incorporated elements of historic Engadine and Tirano architecture, which subsequently became icons of Milanese architecture; the Giò Ponti design features many hydroelectric power plants along the corridor (Valposchiavo, Valtellina, Valchiavenna).

The selection of an economic-managerial model for transportation services requires integration with the settlement and development model. This integration must be considered in relation to the specific characteristics of the context and the diversity of morphologies, employing a multiscalar approach. The criterion of densification in railway stations, applied in metropolitan "cores", is replaced by the

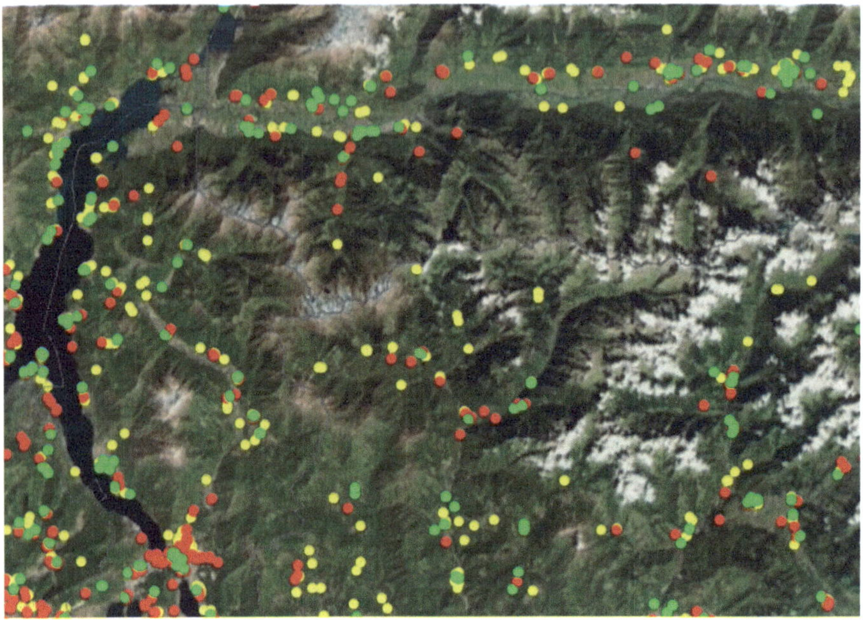

Fig. 2 Heritage concentration along the Lake Como-Valtellina section of the Bernina Corridor (buildings with listing are indicated in red, buildings selected by the Italian Touring Club in yellow, buildings with both of these features in green). Source: GIS Geoportale Regione Lombardia; basemap: Google Earth, Image Landsat/Copernicus 2021

flexibility of articulated and synergic systems in pre-Alpine and mountainous territories. The primary reference is the European CODE 24 research project, which focuses on the development of settlements along the Rhine-Alps infrastructure corridor (Drewello, Scholl, 2016; Otsuka et al., 2017).

The "Slow" Bernina Corridor Between Milan e Zurich

The Sport and Heritage Corridor represents an initiative to identify and promote a transalpine route that would complement the existing Simplon and Gotthard axes. This route would serve as a model for integrated mobility and settlement in mountain context, offering an alternative to the Transit Oriented Development approach.

Such a proposal can be interpreted as the "slow" Bernina axis in comparison to the high-capacity, high-speed Simplon and Gotthard axes. It forms a route from Milan, through Valtellina and the Bernina and Albula passes, to Zurich and Basel, where it rejoins the other two major transalpine backbones touching Milan, the Ceneri-Gotthard-Zimmerberg route and the Simplon-Loetschberg route which both are balancing freight and high-speed passenger travel. These represent three alternative and complementary routes within the trans-European Rhine-Alps corridor

(Fig. 3). They are situated within an evolving interregional and interval cross-border scenario that offers new opportunities for the Valtellina.

It is erroneous to assume that a slow axis implies an unstructured public transport axis. Both Switzerland and Lombardy organize their rail networks according to a frequency that offers users a regular connection, not the fastest possible, but rather the necessary one, connecting the nodes of the system. In peripheral regions of Switzerland, regular connections every 30 min or every hour or even every 2 h between different modes of transportation are already standard practice. Extending the system to the entire corridor allows tourists to flexibly combine different modes of transportation, round trips and stopovers.

This set of transalpine backbones is proposed as a system structure, in a strategic perspective of resilience of the mountainous and pre-alpine cross-border territory, flexibility and diversity of settlements, also to respond to global scenarios of economic crisis and new challenges posed by the experience of public health emergencies. The three routes show different and complementary potentials, while sharing a common criterion of integration between urban and transport planning.

The Gotthard axis is the paradigm of volumetric concentration around major high-speed rail stations, according to a polycentric metropolitan model that densifies in conjunction with land saving and capitalizes on the availability of industrial areas or brownfields close to major nodes. Travel times between the two metropolises of Milan and Zurich will be significantly reduced (from the current 3 hours to about 2.5 h), while Lugano will move closer to Milan (60′), increasingly becoming an integral part of the Milan metropolitan region. The question remains as to how much priority should be given to the speed of the connection between the poles at the level of the macro-region (Milan and Zurich) versus an interpretation of the backbone that is more in the service of polycentrism at the regional level.

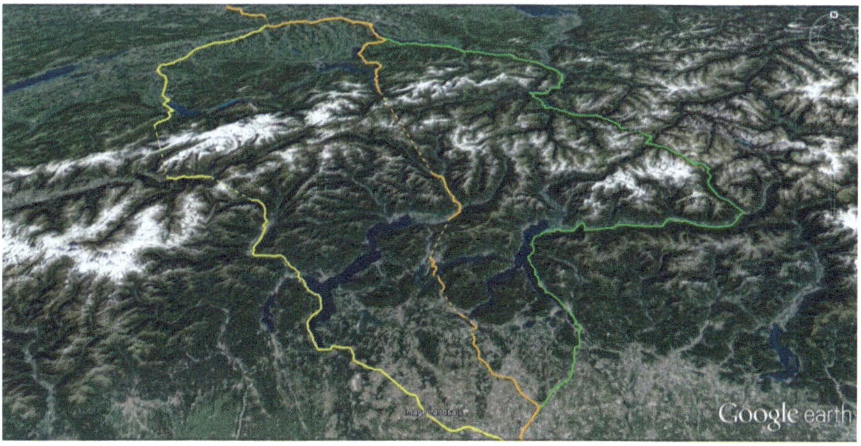

Fig. 3 The 3 transalpine corridors (Simplon, Gotthard, Bernina) connecting Milan and Zurich metropolitan areas. Source: image elaboration by G. Fossa; basemap: Google Earth, Image Landsat/Copernicus 2021

The Simplon axis plays between the freight capacity that supports the logistics hubs of Domodossola, Gallarate and Novara, and the considerable tourist potential of the Monte Rosa, Lake Maggiore and Geneva regions, still unexplored by Italian railways, unlike those in Valais. It is a hybrid corridor between a high-capacity international crossing and a tourist service for the Vaud-Sierre-Ossola-Lake Maggiore area.

In this context, the Bernina Corridor is proposed as the corridor of experiential, slow and sustainable tourism: it should not be seen only from Milan to Valtellina and St. Moritz, limited to the image of the "little red train", but it should be considered in its connection, through the Albula Pass, to Zurich, seamlessly. From this point of view, its potential to enrich the diversity, the interconnection and thus the resilience of the entire Milanese transalpine backbone system is obvious.

Of particular strategic importance in this vision is the hypothesis of a tunnel under the Forcola pass, which would provide a short but crucial route for direct engagement with the Livigno ski area on the Bernina pass.[4] Of similar importance is the long-standing discussion concerning the train connection between Tirano and Edolo, offering continuity to the railway line going up from Brescia via Iseo. The resolution of the mid-lake junction (Varenna-Bellagio-Porlezza) would also permit the implementation of a train-boat sequence, facilitating transportation to Lugano and Locarno. These destinations could then be connected with the "Centovalli" train in the Ossola Valley.

It can be postulated that the systemization of currently disparate sustainable mobility routes may facilitate the establishment of a trans-piedmont connection between the three transalpine routes (Simplon, Gotthard, and Bernina). This could serve to enhance the flexibility and adaptability of the overall system in response to changing circumstances.

In contrast to the model of settlement integration observed in other corridors, the Bernina model does not entail a process of concentration and densification in nodes. Instead, it is characterized by a commitment to preserving the historical system of centers and settlement cores, which are deeply embedded in the complex morphology of the Alpine and pre-alpine valleys. This approach encompasses valley-floor and lake conurbations, and it enhances the multifunctionality of coastal routes, interval connections, and their junction nodes on the valley floor. These nodes often exhibit a rich historical legacy, characterized by the presence of numerous service facilities. The current health emergency demonstrated the value of the multiplicity of small hospital facilities in these territories, which had been regarded as a dispersion of resources and planned for concentration. Instead, they proved to be a source of resilience. In Valtellina, the Chiavenna hospitals and the former Sondalo sanatorium had been transformed into centers for treating patients with the coronavirus, while Morbegno and Sondrio continued to provide regular hospital services.

The Tirano node, situated at an interchange between Italian and Rhaetian railways and positioned centrally along the Adda greenway, has the potential to evolve

[4] The hypothesis is proposed by the current "Regional Territorial Plan for Media Alta Valtellina" by Regione Lombardia.

into the focal point of an interval cultural district. It could emerge as a prominent regional hub situated between Milan and Brescia and with international connections to the Grisons. The downstream stations of the Terrace Route have the potential to become logistic bases that integrate services for inhabitants, tourists, and students, facilitating new lifestyles, work, education, and the development of ecological vehicular mobility (shuttles on unpredictable roads) and on-demand drone services at high altitude.[5] The Colico node can be redesigned by enhancing its synergy with the Pian di Spagna reserve, which can be reinterpreted as a usable park capable of absorbing flooding intensified by climate change. The spatial and temporal frequency of intermodal exchange possibilities and the comfort of such exchange at the nodes (bike and car sharing stations, bike on train and boat, integrated apps and fares, etc.) are conditions for landscape regeneration, in a social and environmental sense.[6]

In synthesis, the proposed corridor (Fig. 4) represents a smart corridor that develops a fine-grained and site-specific resilient polycentric strategy. This strategy is capable of generating counter-cyclical flows, supporting the survival of local mountain communities, ensuring a continuity of living and working, and attracting tourists. Furthermore, it protects the health of travelers (tourists or commuters) through sporting activity and takes care of their psychological well-being through the enjoyment of an identity landscape and a sense of belonging to a living and authentic community.

A Cross-Border Connectivity Approach

The ability to access a linear network of hiking and cycling trails at various entry points is of particular value as it enables users to explore the region in a progressive manner, even on shorter segments of a long trail. This flexibility allows visitors to experience the landscape in a way that aligns with their schedule and interests, thereby fostering a gradual and more in-depth discovery of the area's natural and cultural features. The division of a lengthy trail into multiple sections allows for a more inclusive trail system, which can be accessed by a wider range of users, including families, casual walkers, and cyclists. This approach encourages repeat visits and supports local businesses at each access point, enhancing the overall sustainability and accessibility of the trail system.

The incorporation of existing Alpine routes into the Slow Corridor Milano-Valtellina-Bernina underscores a unified timetable, facilitating seamless transitions at pivotal transportation hubs such as Lecco, Varenna, Colico, Morbegno, and Tirano. These nodes already offer services at 00/30 or 15/45 intervals, but there is a

[5] See chapter Fossa, G., Guenther, F., Paolini, D. "The Smart Terrace Trail as Strategic Regional Project".

[6] The topic has been discussed during the Regional Planning Lab Fossa-Günther at Politecnico di Milano for environmental engineering students, A.Y. 2022–2023.

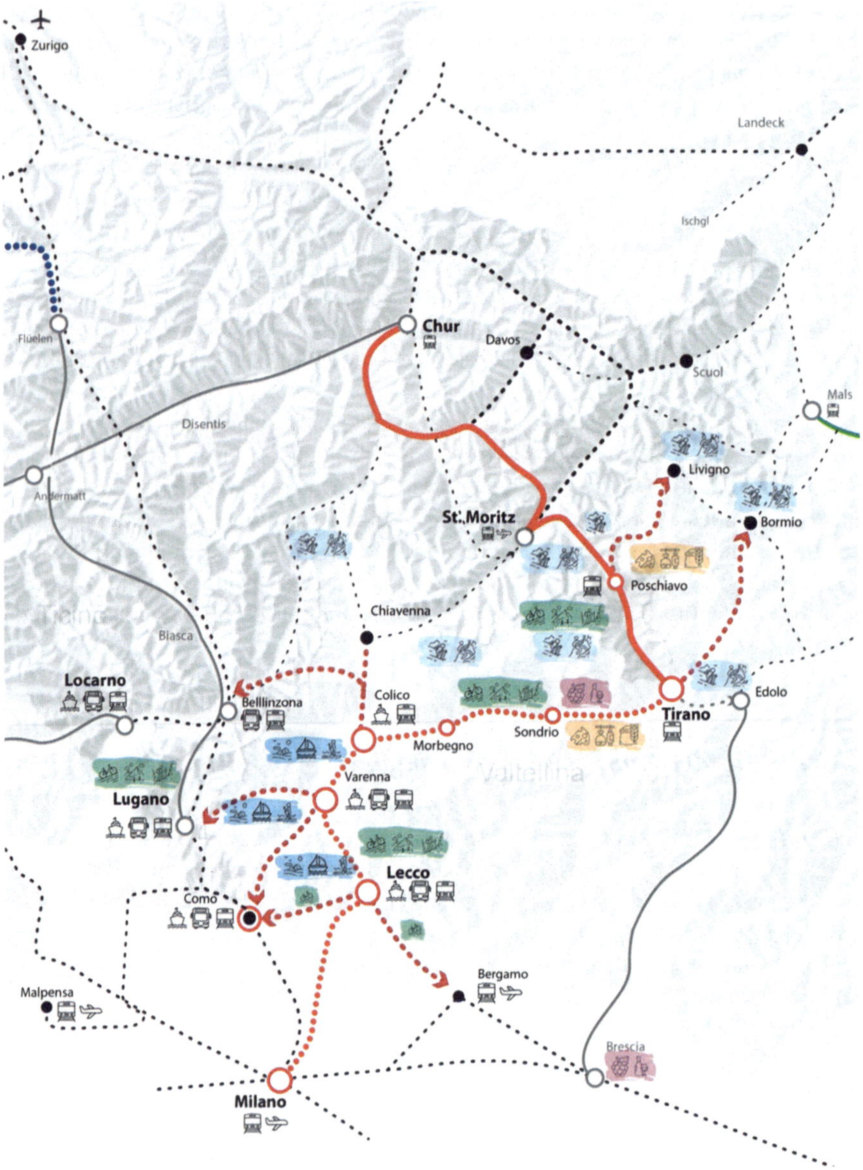

Fig. 4 Sports and Heritage Corridor, integrated in the touristic train network. It is evident the potential for a piedmont touristic axis cross-connecting the N/S corridors of Simplon, Gotthard and Bernina. Source: image by F. Günther

chance for additional alignment, offering travelers effortless transfers between trains, buses, and boats, thereby minimizing waiting times and encouraging greater use of public transport. In our proposal, at key nodes such as Lecco and Varenna, connections to Lake Como boat services and cycling paths could facilitate tourist mobility, enabling easier exploration of the surrounding regions. Similarly, in Colico and Morbegno, the integration of shuttle buses, electric bikes, and train services serves to reinforce the corridor's commitment to sustainability. Tirano, the final interchange point on the route, already provides connections to the Bernina Railway, which leads into Switzerland, but an optimized schedule not only improves travel flow but also supports the broader goals of slow tourism. It ensures that visitors can enjoy a relaxing, eco-friendly journey through the Milano-Valtellina-Bernina Corridor while experiencing the beauty of the Alps and Lake Como regions.

The development of transport platforms along the Corridor could offer a wide range of services and amenities to attract tourists, drawing from successful experiences like the Grand Train Tour Switzerland and regions such as Val Venosta and Val Pusteria in Italy. The primary objective is to enhance convenience, promote sustainability, and enrich the overall travel experience while showcasing local culture, nature, and gastronomy. These hubs would facilitate seamless intermodal connections, enabling convenient transfers between various transportation modes, including trains, buses, boats, and cycling paths. This would allow tourists to easily traverse different regions without the necessity of a personal vehicle. The hubs would also feature bicycle rental services and well-marked cycling paths, facilitating access to popular routes such as the bike Sentiero Valtellina or the trails surrounding Lake Como. The incorporation of e-bike charging stations would enable tourists to explore the surrounding natural areas at their own pace, thereby promoting sustainable tourism practices.

At the aforementioned hubs, tourists would also have the opportunity to partake in local gastronomic experiences, including access to local food markets, wine-tasting rooms, or restaurants offering regional specialties such as Bitto cheese, Bresaola meat, and Valtellina wines. Such hubs could emulate the Swiss approach, whereby transport stations serve as gateways to local gastronomic offerings, thereby enriching the cultural and culinary aspects of the journey. Furthermore, these transportation nodes could serve as locations for tourist information centers, similar to those found in Swiss hubs such as Chur or Interlaken. Such centers could provide guided tours, cultural events, and curated itineraries. Tourists could readily reserve tours to local vineyards, historical sites, or mountain treks directly from the hub, thereby enhancing convenience and supporting local tourism businesses. By integrating these offerings at strategic transport hubs along the corridor, the region would create an appealing, continuous, and sustainable travel experience, encouraging tourists to explore the natural beauty, cultural heritage, and culinary richness of Valtellina and beyond.

Enhancing the Quality of Collective Mobility

With the Gran Train Tour Switzerland, Swiss train operators are proposing a network of touristic train offers, that already provides international tourists with seamless access to Valtellina. The Cross-border Bernina Line promotes a scenic and sustainable way to explore both the Swiss Alps and Valtellina, between Tirano and Swiss cities like St. Moritz and Poschiavo, attracting high-end international tourists. The bus services Palm Express between St. Moritz and Lugano and Bernina Express between Lugano and Tirano complete a touristic offer for mainly Switzerland-based tourists. These offers form only a small part in the much larger offer of the Grand train tour Switzerland which includes also well-established panoramic trains like Glacier Express, the Golden Pass Express and the Gotthard Panorama Express. The Golden Pass Express demonstrates that different railway systems are no impediment for touristic corridors. It offers a flawless transition from Montreux to Interlaken and further to Lucerne, passing through diverse landscapes from lakes to mountainous regions. The train is notable for its variable gauge technology, which allows it to seamlessly switch between narrow gauge and standard gauge tracks without requiring passengers to change trains. This technical innovation enhances the connectivity across different rail networks in Switzerland. The panoramic design of the train allows passengers to fully appreciate the stunning views of the Swiss Alps, vineyards, and forests, making it a popular choice for tourists year-round.

The provision of high-quality tourist experiences in collective mobility is exemplified by the Swiss railways Golden Pass Express and Bernina Express. These examples demonstrate that the appeal of a touristic journey is less dependent on the infrastructure and speed of the journey itself, and more on the quality of the service and the amenities provided. Such trains, with their panoramic windows, comfortable seating, and dedicated services, facilitate an immersive experience for passengers, allowing them to engage fully with the scenic landscapes and cultural atmosphere. For a touristic linear corridor, this model of collective sustainable mobility is optimal, as it permits visitors to traverse the region at a leisurely pace, intertwining with walking, hiking, biking, skiing or watersports activities along the slow corridor at various points of interest along the route, without the environmental impact of cars. This approach is aligned with the concept of slow tourism, wherein the journey itself is regarded as a valuable aspect of the travel experience. It fosters meaningful connections with both the local environment and community. By emphasizing sustainable, high-quality services, linear corridors enhance the accessibility and attractiveness of the region. They integrate transportation as a core aspect of the overall tourist offering.

The development of comprehensive touristic routes combined with high-quality, sustainable collective mobility is essential for creating a truly immersive experience for visitors. The proposed Sport & Heritage Corridor in Valtellina illustrates how the journey itself becomes a core component of the tourist offering, enriching travellers' engagement with the landscape, culture, and activities. Inspired by successful models like the Swiss Bernina Express, this corridor emphasizes quality services, scenic routes, and multimodal accessibility over speed, thus enhancing both the travel experience and environmental sustainability. This approach offers travellers flexible access points and a seamless transition between different modes of mobility—whether by train, boat, or bike—without the need for car travel. By establishing a corridor where moving through the landscape is part of the destination itself, the region not only attracts diverse visitor profiles but also supports year-round tourism.

This sustainable, experience-based infrastructure serves as a replicable model for other transalpine regions, fostering regional resilience and preserving cultural heritage through thoughtful, integrative mobility solutions.

A New Narrative for the Milan-Cortina Connection

At last, a landscape synthesis has emerged, namely the concept of the "Sports & Heritage Corridor". The definition of the "slow" transalpine corridor through the Bernina Pass is based on historical precedents. A brochure from 1912,[7] the year the Bernina railway line was inaugurated, illustrates the tourist orientation of the new line, which aimed to connect major urban centers (Vienna, Berlin, London and Paris) with Milan, while also serving the Alpine tourist resorts of Zermatt, Innsbruck and maritime destinations such as Venice and Nice. A number of popular tourist offerings are based on the rail network, including the Glacier Express (St. Moritz-Zermatt line & Bernina Express).

Now this idea responds to the need for a new narrative of the Milan-Cortina Olympic binomial. The Po Valley axis via Venice, evoking the Serenissima Republic, represents not only a historical route but also a means of repositioning Milan at the center of the Alpine arc. This contemporary reinterpretation of historic crossings, particularly the "salt route" connecting Milan to Salzburg via Valtellina, and the "wine route" between Valtellina and the N/E pre-Alps,[8] allows the vision of the Olympic landscape to be unified. By systematically organizing the venues of Olympic events distributed along the central-eastern arc of the Alps, this corridor could serve as a platform for regional branding on a global scale.

The proposed corridor can thus become Milan's privileged graft into the settlement system of the Alpine valleys, a cross-border system articulated on the watersheds.

This "passage to the North-East", which traverses the Adda and Inn basins, connects Milan to Cortina across the Alps. It also links the Bernina route to the Brenner axis, effectively closing the loop of the "Po Valley" connection via Venice.

---

[7] Tourism promotional brochure "Bernina-Bahn", 1912 (Rhatische-Bahn CH).

[8] See chapter Fossa, G., Guenther, F., Paolini, D. "The Smart Terrace Trail as Strategic Regional Project".

This imagery revisits the historic Lombardy-Veneto region, incorporating the Swiss valleys into a cross-border vision that traverses the Alpine watershed. This vision may be regarded as both a metaphor and an integrated system of sustainable mobility. Furthermore, it may be seen to reinstate the Valtellina region as a central component of the Alpine macro-region.

Acknowledgements Authors would like to thank Roberto Laffi, Director of Territory and Green Systems, Regione Lombardia.

References

ARE, Ufficio Federale dello Sviluppo Territoriale. (2024). *Progetto territoriale Svizzera*. Retrieved November 8, 2024, from https://www.are.admin.ch/are/it/home/raumentwicklung-und-raumplanung/strategie-und-planung/raumkonzept-schweiz.html

Drewello, H., & Scholl, B. (2016). Integrated spatial and transport infrastructure development. In *The case of the European North-South Corridor Rotterdam-Genoa*. Springer.

Fondazione Milano-Cortina 2026. (2024). *Programma per la realizzazione dei giochi olimpici e paralimpici Milano-Cortina 2026*. Retrieved November 8, 2024, from https://milanocortina2026.olympics.com/s3fs-public/documents/2024-06/Allegato%201-Quadro%20di%20riferimento%20programmatico.pdf?VersionId=qlpHWnwMRb3NOjyD.X8LJmwnlCBq3776

Otsuka, N., Günther, F., Tosoni, I., & Braun, C. (2017). Developing trans-European railway corridors: Lessons from the Rhine-Alpine Corridor. *Case Studies on Transport Policy, 5*(4), 527–536. https://doi.org/10.1016/j.cstp.2017.06.005

Regione Lombardia. (2021). *Revisione del PTR Piano Territoriale Regionale (adopted document)*. Retrieved November 8, 2024, from https://www.regione.lombardia.it/wps/portal/istituzionale/HP/DettaglioRedazionale/servizi-e-informazioni/Enti-e-Operatori/territorio/pianificazione-regionale/revisione-ptr-piano-territoriale-regionale/revisione-ptr-piano-territoriale-regionale

UFCOM. (2023). *Il Consiglio federale vuole un accesso Internet veloce per tutta la Svizzera*. Retrieved November 8, 2024, from https://www.admin.ch/gov/it/pagina-iniziale/documentazione/comunicati-stampa/comunicati-stampa-consiglio-federale.msg-id-99274.html

Remote Working Strategies as Regional Development Frameworks

New Geographies of Work: "South Working" to Retain and Attract Human Capital

Elena Militello

Introduction

The sudden onset of the Covid-19 pandemic in early 2020 led to widespread questioning of the spatial and temporal paradigms of work in the twenty-first century. While some have argued for an unprecedented overcoming of the built-in frailness of traditional work settings in highly dense urban contexts, others have been more cautious about celebrating a new era of work and highlighted the ephemerality of the pandemic, as a temporally confined event.

Despite the multifaceted positions, the reactions to the health crisis undeniably inspired transversal economic, social, and cultural changes, notably reshaping of the geographies of work. Many highly skilled professionals faced an unexpected shift in their daily habits by having to cope with lockdowns and being able to work remotely for significant periods. The pandemic thus accelerated digitalization and required multilevel digital skills improvements.

While pandemic-induced transformations occurred throughout the world, their impacts varied based on local technological and social preparedness. In Europe, for example, the impacts were particularly significant in Southern countries, which have historically experienced wide territorial, social, and economic disparities. This set of pre-existing fragilities inevitably exposed whole areas of Europe to higher levels of uncertainty and socio-economic challenges. These gaps manifest both internally (within their own countries – for example between Southern regions and Northern ones in Italy – and between non-metropolitan areas and big cities) and externally (between countries with weaker economic performances and those with more prosperous economies).

E. Militello (✉)
Minda de Gunzburg Center for European Studies at Harvard University, Harvard, MA, USA
e-mail: elena.militello@southworking.org

G. Fossa et al. (eds.), *Small-Town Renaissance*, Research for Development,
https://doi.org/10.1007/978-3-031-89786-3_6

They encompass both macroeconomic indicators (*e.g.*, GDP *per capita*; general, youth and women unemployment; rate of NEETs - "Not in Education, Employment or Training") and social equity (*e.g.*, infrastructural deficits; limited healthcare access; non-universality of the right to preschool education). Historically, these regions also faced demographic shrinkage, stark migration flows and, in recent years, brain drain, both intranational (towards other regions of the same country) and international (towards other countries). While these economic and territorial settings have traditionally shaped the geography of work at least since the mid-twentieth century, the pandemic and the technological are now driving a radical spatial shift.

During the pandemic, these economically and demographically impoverished regions saw an influx of remote workers (often highly skilled professionals), due to returning emigrants and the perceived actual quality of life and lower living costs. While the number of workers who have seized this shift to remote working to move from big cities to less dense areas has been empirically investigated in the United States, its European remains an object of research.

Work-from-anywhere, remote working, and agile working have been examined from various disciplinary perspectives, from sociology to psychology, from law to business management, and from urban planning to economics. These innovative ways of working present both advantages and disadvantages. Yet, a comprehensive assessment of the two is still under debate.

Southern European countries, like Italy, Spain, Portugal, and Greece, represent ideal case studies as to the size of the phenomenon and the feasibility of potential policy proposals to reduce the existing gaps: their economies are sluggish; their territorial gaps are profound; they are particularly hit by brain drain; they are attracting high flows of remote workers.

In Italy, remote workers relocating to Southern and non-metropolitan areas (often called "Inner Areas") while employed by companies in Northern cities or abroad, soon became known as a new phenomenon worth identified through a neologism: "South Working" (Treccani, 2020). By chronicling the normative, economic, and spatial conditions in which the "South Working" phenomenon has occurred, this chapter aims to understand whether this phenomenon plays any significant role in contributing to (1) bridging the existing divides between different areas of Italy, (2) contrasting the brain drain flows, and (3) improving the social innovation capital in these regions. It explores what its constituencies and prerequisites are, and what public policies could be enacted to stimulate the relocation of existing jobs (or, even, the creation of new ones) in historically marginalized and economically deprived areas, focusing on periods of remote working conducted from coworking spaces defined as "community hubs".

Specifically, this research employs a unique choreography of spatial, quantitative, and qualitative datasets: data from an extensive user-oriented survey conducted between 2020 and 2022 (2594 responses) are analyzed to extract the profile of workers who are more likely to work remotely from Southern Italy; data from a coworking-targeted survey highlight the spatial distribution of 230 spaces suitable for remote working; finally, open-ended survey entries from 18 young professionals who experienced periods of "South Working" add relevant nuances on the individual preferences on how, where, and for what reasons they conducted these extended periods, which will be analyzed through sentiment analysis and word clouds in a separate research project.

While the "South Working" phenomenon has garnered extensive press coverage through anecdotal evidence and powerful storytelling, there is limited knowledge on the profiling of workers interested in working remotely from these regions and in investing their energies to revamp local communities. In light of these social transformations and remote working's possibilities, the European Union's territorial cohesion policies require an in-depth reconsideration to meet the contemporary challenges of emerging new geographies of work.

Background Analysis

The Socioeconomic Context Prior to the Pandemic Outburst in Southern Europe

Analyzing "South Working" requires an understanding of Southern Europe's pre-pandemic socioeconomic conditions, *i.e.* the context in which the phenomenon developed in early 2020.

In Italy, the 2020 health crisis exacerbated a gloomy pre-existing scenario, mostly shared by Southern regions ("Meridione" or "Mezzogiorno") (Bianchi et al., 2019) and Italian Inner Areas (Lucatelli et al., 2019).

Since its unification in 1861, Italy has always experienced significative regional disparities: on the one hand, when it comes to the "Southern Question" ("questione meridionale") (Schneider, 1998; Salvemini, 1955; Gramsci, 1991), i.e. the divide between "two Italies", a Northern one and a Southern one; on the other hand, the divide between urban centers and non-metropolitan areas and smaller towns, gathered under the label of "Inner Areas" ("Aree Interne"). In recent years, the former Italian Ministry of Territorial Cohesion (now Department of Cohesion) mapped the whole country on the basis of the availability (or lack) of services and infrastructures, specifically transportation, education, and health (Fig. 1). As a result of this in-depth analysis of the Italian landscape kickstarted by former Minister Fabrizio Barca in the early 2010s, it has been proven that Inner Areas encompass nearly 4,000 municipalities (about half of the total number of municipal governments). These municipalities cover 58.8% of the national landmass and house about 13.4 million people (22.7% of the 2021 resident population).

A similar demographic distribution, between "empty" regions and "full" or "dense" ones, acts as a common denominator across several areas of Southern Europe. It also characterizes Spain's inner areas, except for its capital city Madrid (Encarnación, 2022), it is mirrored by the concentration of population in Greece around its main cities (Athens or Thessaloniki) and it is similarly shared by Portugal,

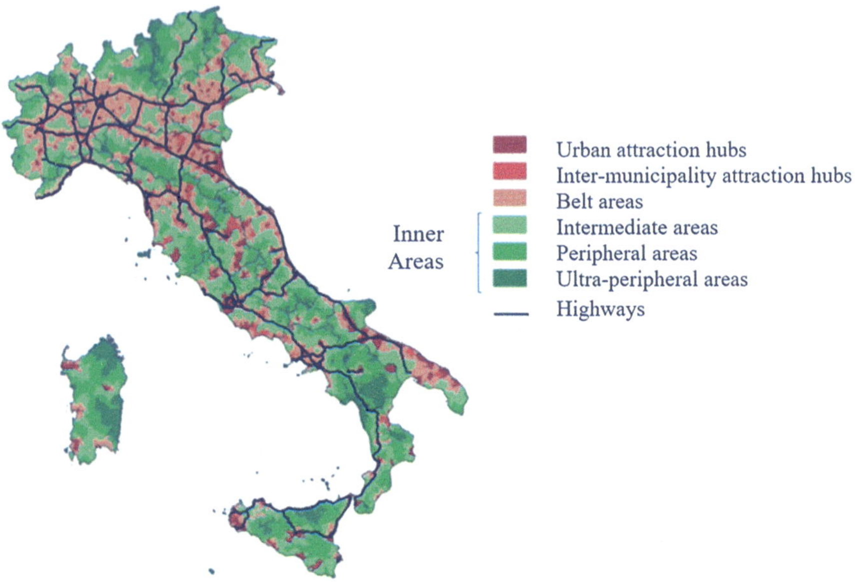

Fig. 1 Inner Areas ("Aree Interne") in Italy. Source: Italian Agency for Territorial Cohesion (Agenzia per la Coesione Territoriale); data gathered from the Italian Ministries of Health and Education and from the Italian State Railways ("Ferrovie dello Stato") – English Labels by Author

especially in its coastal areas and around its main attraction hub, the capital city of Lisbon (Fig. 2).

Macroeconomically, Southern Italy faced a slight recession (−0.2% of GDP) already in 2019, while Northern Italy, albeit essentially stagnant, grew by 0.3% (SVIMEZ, 2019). Southern Italy lagged significantly behind the European Union ("EU") average in all indicators (infrastructure, health, transport, education, unemployment, gender equity, poverty, and legality).

In 2017, the GDP *per capita* of Southern regions in Portugal, Spain, Italy and Greece, was in the lowest bin (below 75% of the EU average).

As to the labor market, the very low levels of employment rates were already causing concerns in 2019, especially for women and youths, because of their employment rates at around 33% (EUROSTAT, 2019), with a very high risk of falling into the category of NEETs (Openpolis-Con i Bambini, 2019). This trend seems to be stable according to data from 2021 (Chiellino, 2023) and 2022 (SVIMEZ, 2022). Sicily, specifically, was in 2021 the EU-27 region with the highest youth unemployment rates overall (40.1% in the age group 15–29 with an already alarming general unemployment rate of 18.7%) (Fig. 3).

This dire scenario has historically gone hand in hand with issues in accessing essential public services, which, according to the Italian Constitution, should be granted to all citizens. These gaps cause what became known as "citizenship gaps" ("*divari di cittadinanza*") (Bianchi & Fraschilla, 2020), hindering the possibility of attending full-time primary schools, access to healthcare and, more generally, the

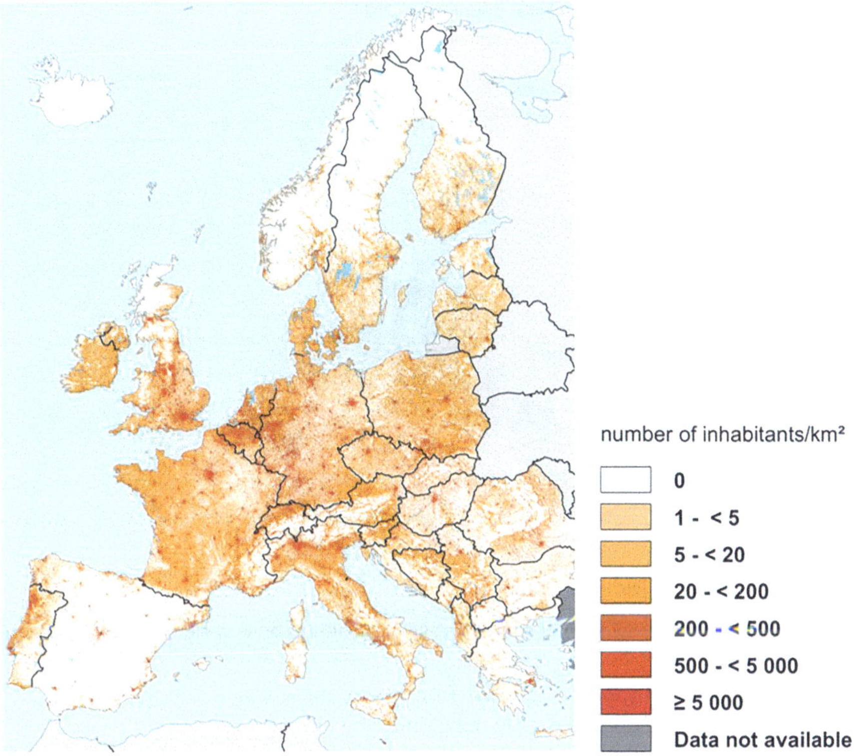

Fig. 2 Population density, 2011. Source: JRC, Eurostat, GEOSTAT Population Grid 2011

persistence and worsening of territorial and infrastructural gaps, with declining public capital expenditure in marginalized regions and EU funds replacing rather than supplementing domestic resources (SVIMEZ, 2019).

Southern regions have also historically suffered from demographic shrinkage and stark migration flows. In recent years, these flows have turned into a "brain drain", both intranational (towards other regions of the same country) and international (towards other countries, both in the EU and elsewhere). Brain drain is broadly defined as the phenomenon of intellectual migration flows of (often highly) skilled and educated young professionals to other regions, for research or employment opportunities. This leaves the areas of departure with a reduced supply of skilled people not only in research but also in public and private services. Within the EU, internal migrations from one Member State to another are referred to as "mobility", not migration (Recchi & Triandafyllidou, 2010).

In Italy, the phenomenon of migrations from the South and Inner Areas towards large cities of the North and abroad has been compounded by an increasing new trend of migrations, so-called intellectual migrations, of university students and/or highly skilled workers (Sanfilippo, 2015; Vecchione, 2017). This phenomenon has been certified by the Italian "Corte dei Conti" (Court of Auditors), which, in 2021,

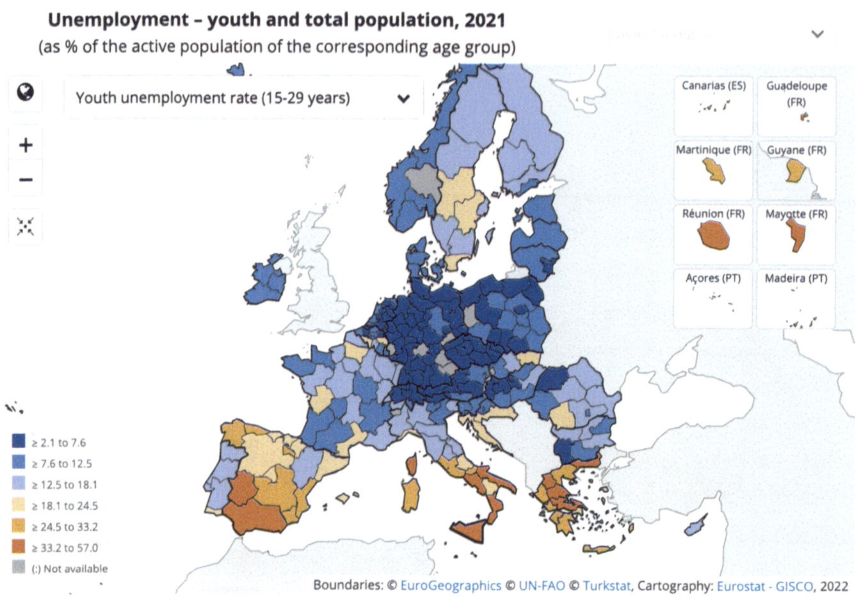

Fig. 3 Cartography of youth and total unemployment in Europe. Source: Eurostat—GISCO

calculated that, within the 2013–2021 timeframe, there was a +41.3% increase of graduates emigrating abroad (Corte dei Conti, 2021).

While these statistics are already quite alarming, they are most likely an underestimation of the magnitude of this phenomenon. They do not consider those who have moved their residence to other Italian provinces, nor the many who only move their domicile and not their residence, which makes it extremely difficult to extract reliable data on relocation trends. A new line of research tests the possibility to use telecommunication ("TLC") data to show patterns of mobility more accurately (Mariotti et al., 2022c; Pimpinella et al., 2022).

The difference between subjects who voluntarily decide to leave their hometowns ("opportunity migrants") and those who feel forced to do so without wanting to ("necessity migrants") is blurred for residents of areas featuring a lack of alternative opportunities for personal development (Sonzogno et al., 2022).

Many studies show how geographic mobility experienced by workers because of intellectual migrations and brain drain entails uncountable psychological costs and consequences due to the natural will to maintain social attachment to a place, combined with the feeling of despair or impossibility to move back (Choudhury & Kwon, 2018).

Similar demographic hemorrhages occurred after the 2008 crisis in Spain (Nelson, 2015), Portugal (Ganga et al., 2018), and Greece, where more than 500.000 people left the country over the past decade, out of a population of around 11 million (Pratsinakis, 2022).

If we look at the forecasts as to population developments between 2021 and 2050, almost every region of the countries in Southern Europe is expected to lose significant chunks of the population (Fig. 4), with a population decline not only linked to lower fertility rates but also to a centripetal force towards central Europe.

Additionally, in the upcoming decades, Southern Europe is also expected to be hit more harshly and directly by the effects of climate change, which might increase Northbound migration flows because of land erosion and sea-level rise.

The Spread of Remote Working

Despite discouraging preexisting conditions, this section focuses on how Southern regions have also welcomed, during the pandemic, high numbers of remote workers, often highly skilled professionals.

For many years, before the pandemic, the apparently unstoppable rise of the tech industry, combined with the ever-decreasing importance of the primary and secondary sectors in the economy, led many policymakers and scholars to invest in the concept of creativity concentration in major urban hubs as sites of positive community impact (Florida, 2002). Big cities, even metropoles, enjoyed a season of triumph and were believed to be able to magnify the progress of humanity (Glaeser, 2011). Those cities that managed to attract clusters of innovative firms from cutting-edge sectors, i.e. those characterized by steadily increasing innovation levels and by

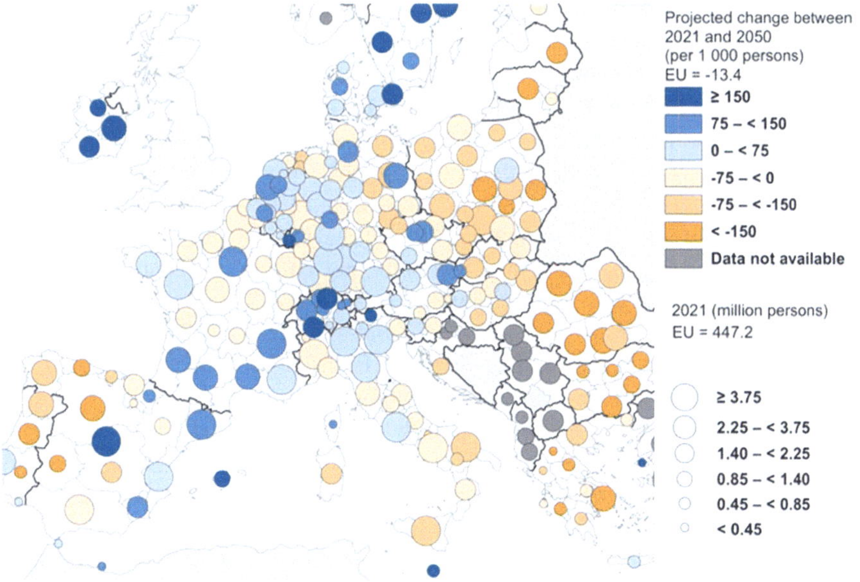

Fig. 4 Population developments, 2021–2050, by NUTS2 regions. Source: Eurostat

the need for a highly educated workforce, even in the most expensive areas, were seen as "brain hubs" (Moretti, 2012). They would engender positive spillovers and a virtuous circle of attraction of the "best workers" to the "best jobs", in a flurry of technological optimism and competitive success (Choudhury et al., 2018). Those same workers, however, paid the price of higher rents or real estate prices and often unbearable costs of living, often far away from their inner social circles. Their presence, en masse, also impacted the city environment in which they were grafted, by creating a deepening of injustices and gaps with the population employed in lower-paying jobs. Besides, the myth of eternal progress clashed for the umpteenth time in history with unforeseen events like the pandemic, the subsequent economic crisis, the tech industry layoffs, and the breakthroughs in generative artificial intelligence that worried even tech leaders (Mohammad et al., 2023).

In the meantime, however, reflections on the "future of work" started fostering the idea of transforming, for some types of employment, the traditional forms of work linked to the mere physical presence at the employer's premises during working hours. This paved the way to a more "agile" way of working, based on greater flexibility and above all on the definition of objectives, cycles, and phases. There are already many management studies that underline the potential benefits of agile working for the employer, in terms of increased productivity, improvement of workers' digital skills and their motivation, reduction of overtime and low productivity, cost optimization, and effects on reputation and corporate social responsibility (Neeley, 2021).

These reflections were mostly carried out in Anglo-Saxon or Northern European contexts, where performance indicators (so-called KPIs, key performance indicators) were already more widespread, and attention to the issue of reconciling personal and professional life was considered a critical goal (Angelici & Profeta, 2020). International bodies dealing with labor had started, well before the pandemic, talking about "workplace flexibility" as a perspective for balancing personal and professional life (OECD, 2016; Eurofound-ILO, 2017).

To trace back the remote origins of those proposals, we must go back to a 1973 UK intuition (curiously in the middle of yet another energy crisis), with the invention of the neologism "telecommuting" (Nilles et al., 1976), as a form of telecommunications surrogate for commuting. There, the idea of physical spaces with shared workspaces spread all over the country, not only in city centers, emerged, through the use of the term "coworking".

The category of "remote work" or "work-from-anywhere" ("WFA") (Choudhury, 2020; Choudhury et al., 2021) includes both forms of work that are carried out exclusively at home and are characterized by time and space constraints ("telework" or "telecommuting"), and those organized by phases, cycles, and objectives carried out in different locations ("agile work") (Junker et al., 2022). Agile methodologies derived from the world of project management became known in Italian, with a pseudo anglicism: "smart working" (Bianchi & Vecchione, 2021).

Obviously, not every job can be performed remotely, and the chance and positive effects of remote working and work flexibility are positively correlated to the size of the company (the bigger the company, the easier to work using KPIs) and to the workers' specialization in knowledge-intensive sectors (Sostero et al., 2020).

The growth of remote working, therefore, produces effects on the company organization and productivity, but also on the types of workspaces and the geography of work, with significant implications on the center-periphery relationships, between urban and non-metropolitan areas (Choudhury et al., 2021; Felstead & Henseke, 2017).

At the peak of the Covid-19 pandemic, the need to foster "physical distancing" to minimize the chances of contagion and transmission of the virus raised the need to also change the way individuals used to work all over the globe: many private workers and professionals, as well as many public employees, were suddenly urged to work remotely (Mariotti et al., 2023).

In the past five years, it has clearly emerged how much and in which ways remote working has "emptied" cities, making peri-urban and peripheral areas more attractive (Bellandi et al., 2021) and spearheading a renewed interest in livable (walkable and bikeable) neighborhoods, spreading the heavily debated concept of the "15-minute" city (Mariotti et al., 2022c). In fact, the pandemic has weakened large urban agglomerations, since aspects such as high population density, traffic congestion, and pollution have become aspects not only undesirable, but also at higher risk as to the spread of infections (Casti & Adobati, 2021; Musolino & Rizzi, 2020) and, more recently, because of inflated costs of living and unbearable rents (Cascaldi-Garcia et al., 2023).

At the same time, policies adopted to contain the spread of the virus, including remote working, strongly reduced the urban gravitation of workers and the relative demand for goods in city centers, while they increased interest in less dense areas, with lower traffic rates, where people could live and work in the same neighborhood, reducing commuting needs (Croce & Scicchitano, 2022).

Consequently, the pandemic has highlighted the complementarity between centers and peripheries and the need for analyses and policies to look at their joint dynamics, while focusing on the improvement of living conditions for all (Bellandi et al., 2021). In recent years, some types of jobs seemed to be able to move closer to workers and not vice versa. Companies are opening offices in areas close to the workers' residential areas and companies operating in real estate are seeing an increase in demand for hybrid spaces in peri-urban and peripheral areas with good accessibility ("near working") (Mariotti et al., 2023).

Even some of the pioneers of the triumphant city model started questioning its sustainability vis-à-vis the remote working phenomenon and the new needs of highly skilled workers (Florida & Ozimek, 2021). The potential strategic application of remote working as a tool to reduce pollution and improve air quality by reducing unnecessary commuting started being investigated too (Badia et al., 2021).

Some communities in the US Midwestern States started seeing it as a potential tool for their rejuvenation even before the pandemic, for example through the Rural Innovation Initiative (Sisson, 2019) and some thereafter, like Tulsa, Oklahoma (Choudhury, 2022a; Padilla, 2022).

Other countries managed to reverse historical emigration trends (and even activate a "brain gain") through temporary residency and work permit programs, thus allowing people with foreign jobs to live locally, such as Serbia, Montenegro, and North Macedonia in the Western Balkans (Leitner, 2021).

However, remote working was too often adopted in the emergency phase without adequate preparation both by employees and by the top management, and without considering the need to balance the quality of life of local communities with the effects of remote workers. Furthermore, the absence of suitable shared workspaces and a legal system guaranteeing the so-called "right to disconnect", as well as the definition of boundaries between personal and professional life, led to "double and triple workloads", especially for women workers, in households with children and/or relatives for whom the worker is the main caregiver (Angelici & Profeta, 2020).

Research on Italy, France, and Spain conducted by the Joint Research Centre (JRC) of the EU Commission highlighted the effects of remote working during the pandemic, suggesting that it can represent a positive model for the future, provided that the risks of community fragmentation and isolation are minimized, for example through the recourse to coworking spaces (Fana et al., 2020).

Understanding the number of workers who have seized this shift to remote working to move from big cities to less dense areas has been the subject of empirical investigations in the United States, and, to a lesser degree, in Europe.

Between 2019 and 2021, the number of people working primarily from home increased from 5.7% to 17.95% of all workers in the United States and, in Europe, according to some estimates from 14.6% to 24.4% (Murillo Chávarro, 2023); according to Eurofound, the number of teleworkers rose from 11% to 40% across Europe only in 2020 (Eurofound, 2021). Furthermore, the World Economic Forum estimated that, in the USA, the massive use of remote working has led to an increase in labor productivity of 4.6%.

In Italy, the research center "Osservatorio Smart Working" of the *Politecnico di Milano* estimated that the number of workers who have experienced agile remote working has risen from 570,000 in 2019 to approximately 6.58 million during the most acute phase of the Covid-19 emergency in 2020 (Osservatorio Smart Working, 2020).

Even before the pandemic, the number of workers who had experimented with forms of "agile work" (remotely even for just 1 day a week), was growing after the approval of a law in 2017 (see below, next section), reaching 570,000 units in 2019 (Osservatorio Smart Working, 2019).

This process, already underway before the pandemic, has certainly accelerated sharply in the last 3 years, up to a peak of around eight million teleworkers during the first lockdown (Corso, 2020).

In fact, Italy has moved up in the European rankings as to the percentage of people working from home, almost approaching Northern Europe levels during the pandemic. While in 2019 only about 8% of workers worked from home (in the penultimate position in the European ranking on the use of remote working), during the 2020 lockdown about 40% of workers started working from home (Sostero et al., 2020). Even after the end of the emergency, the "new normal" did not go back to the

570,000 "agile workers" in 2019, but estimates stand at around 3.6 million in 2022, with prospects for further growth in 2023 (Balabio, 2022).

A study by PwC-Italy estimates that if all tasks that can potentially be performed remotely were carried out in that way this would give the Italian GDP a boost of 1.2% (PWC-Italy Research Center, 2021).

Remote working, on the one hand, has overturned the paradigm by showing how a new development model is possible; on the other, it has highlighted all the territorial gaps linked to digitalization (so-called digital divide): ultra-broadband Internet reaches only 25% of the Italian population in comparison to a European average of 60%, thus depriving some citizens of the possibility of working and studying efficiently (Cersosimo & Donzelli, 2020).

According to the Italian National Institute for the Analysis of Public Policies (INAPP), economic activities that are most suitable for working remotely, with a high rate of so-called "teleworkability", are professional, scientific, technical activities; finance and insurance; professional services; public administration (Barbieri et al., 2020).

According to the Italian National Institute of Statistics ISTAT (2020), around 8.2 million employees could work remotely (about 36% of the total), mainly in the sectors of ICT, financial and insurance activities, and services provided by the government and other public administrations (Bianchi & Vecchione, 2021; Giuzio & Rizzica, 2021).

As to the spatial distribution of remote workers, recent studies have shown an outbound flow of workers from large cities in favor of suburban and peripheral areas in Italy: mostly, these are commuters who have started working remotely, at home (so-called "near working") (Mariotti et al., 2023, especially for the metropolitan city of Milan). Furthermore, of interest is the phenomenon known as "South Working", i.e. working remotely from the South and from Inner Areas for companies located in the North, on which the case study will focus.

The Existing Legal Framework

The wide territorial gaps still existing within Europe represent a conceptual and legal problem within the EU as the notion of cohesion is one of the founding principles of the Union, enshrined in art. 174–178 of the Treaty on the Functioning of the EU (TFEU). The term cohesion includes economic, social, and territorial aspects, aiming at achieving an "overall harmonious development" of the Union, with the goal of decreasing the divides in the levels of development of the different regions, especially the least developed ones.

This principle is often mirrored in national higher laws or Constitutions, with a vision to grant similar levels of rights to all people under its jurisdictions. In Italy, the principle is recognized in art. 119, para. 5 of the Constitution.

In the aftermath of the COVID-19 pandemic, the EU approved a comprehensive plan of €750 billion for the period 2021–2026, with the intention to help EU Member

States recover from the pandemic-induced crisis. This plan is broadly known as "EU Recovery Fund" or "Next generation EU". In a nutshell, the plan's goal is to reduce the existing cohesion gaps among European regions, and it is upon that basis that the available resources have been allocated among Member States. On those grounds, Italy was promised €68.9 billion in grants and €122.6 billion in loans (to be added to €42 billion in ordinary cohesion funds). These unprecedented resources are the backbone of an ambitious national plan, known as "PNRR" (National Plan of Recovery and Resilience). Per EU regulations, each country must abide by a tight timeframe to accomplish the promised results, targets, and milestones. In Italy, recent political turmoil, frequent changes of leadership, and increasing inflation have led to serious doubts about the timely implementation of its PNRR (Speciale, 2023).

As to the specific topic of this chapter, the possibility to perform remote work depends on its legal feasibility, *i.e.* if workers are allowed to work without being present in the same location as their offices or employers. Understanding the idiosyncrasies of remote working and its potential for the future requires a more in-depth understanding of the legal framework governing this opportunity, including taxation and immigration (visas). On this last topic, in fact, many EU countries have introduced or are introducing specific "digital nomad" visas to allow non-EU citizens to reside temporarily in the EU and work remotely, provided that they can prove yearly earnings over a certain threshold (usually much higher than the average local income) (Rodriguez, 2022). As usual, the law has proven to be quite late on societal developments, and especially so in an emergency setting where workers acted autonomously before norms were changed. However, in some countries, there have been innovative legislative initiatives to regulate agile forms of work and telecommuting.

In Italy, the Parliament approved a specific regulation already in 2018, Law n. 81/2017. In its art. 18, it introduced the concept of "agile work", with a very advanced approach. The introduction to the rationale of the provision highlights the dual objective of "increasing competitiveness" and "facilitating the reconciliation of life and work", using the verb "promote". "Agile work" is defined as a "method of carrying out an employment relationship established by voluntary agreement between the parties, by organizing one's work by phases, cycles, and objectives, and without constraints of having to work during specific times or in a specific place, with the possible use of technological tools for carrying out the work activity". Finally, it should be underlined that in these contracts there is no "fixed workplace", as the service can be rendered both inside and outside the company premises, while the worker is protected from exploitation risks. Indeed, the law reiterates the applicability of the maximum limits on the duration of working hours, both daily and weekly, deriving from the law and from collective bargaining (anticipating the later discussion on the "right to disconnection", but without providing explicit tools to measure working overtime). In the past 2 years, there have been several bills presented in Parliament to amend this law and adapt it to the new needs after the pandemic, including some aimed at restricting the applicability of the norm and many others at expanding its reach. Additionally, a law to include "digital nomad visa" was approved in March 2022 but was only implemented 2 years later, as it required a second-degree ministerial regulation of its prerequisites.

Apart from several emergency provisions, Greece reformed its whole corpus of labor laws in 2021, with a major reform that also included provisions on remote working, amidst several others on caregivers' rights, the "right to disconnect", violence and harassment in the workplace, and the platform economy (Botsari, 2022). Spain, in turn, timely approved a new statute on remote working (*teletrabajo*) in October 2020; and Portugal amended its labor law framework to change remote working provisions, effective January 2022 (Leitão, 2022). As to visas, Spain approved a digital nomad visa through a reform known as "Startup Act" in November 2022, with an income threshold double the minimum wage, and applications are already open (Frost, 2023). Portugal made an important step back from its notorious "golden visa" program for people investing in real estate, which was interrupted in February 2023, after attracting numerous critics due to its unavoidable gentrification effects and pushback from local communities (Kaysen, 2023). Instead, it did approve a new program specifically tailored for digital nomads, which started in October 2022 (Elton, 2023).

This proliferation of legislative innovations in fields related to remote working all over Southern Europe is a symptom of a growing interest in the subject even from a legal perspective, with the awareness that some regulation will be needed to avoid potential pitfalls in terms of workers' exploitation and cost of living and of housing.

The Case Study of South Working in Italy

The health emergency and the use of remote working have provided many workers with the opportunity to leave the bigger cities of Northern Italy to return, at least for some periods each month or year, to the marginalized areas of the Southern regions and Inner Areas.

In this context, the "South Working" phenomenon emerged, at the hands of the non-profit organization "South Working - Lavorare dal Sud"® ("SW-LdS"). From its website, it is "a social promotion project that stimulates and studies the phenomenon of remote agile working from a different location than that of the employer or the company, with the aim of promoting the social and economic development of historically marginalized areas".

South Working represents a new way of thinking about work even beyond the end of the health emergency. This model aims to guarantee workers an adequate balance between private life and professional life and aims to restore social and economic value to the territories of the South and Inner Areas (Bubola, 2021; Ghiglione & Romei, 2021).

South Working can be defined as: "periods of remote (agile) work from the South and from Inner Areas in coworking spaces" (Militello & Mirabile, 2020). These coworking spaces operate as "community hubs", where workers can "give back" to the local community and learn from them in return (Mariotti et al., 2022b). The project is based on flexible working methods and voluntariness. It involves workers that are mainly based in the largest urban districts located in the Center (mainly Rome) and in the North (mainly Milan) of the country, as well as, more generally, workers abroad (SVIMEZ, 2020). The phenomenon, understood more generally as remote work from the South for the North, is attracting interest not only from researchers but also from political decision-makers. They see in it a potential development for the South, generating "geographical relocations", at least for periods each month or each year, going beyond even the traditional policies linked to territorial tourist attractiveness (Cracolici & Nijkamp, 2009) and undermining intellectual migrations (Recchi et al., 2016; Dotti et al., 2013; Nifo & Vecchione, 2014) and the low competitiveness of marginal areas lacking an entrepreneurial class (Musolino et al., 2021). The goal is to move beyond both the attractiveness rhetoric leading to overtourism and extractive tourism (Giambona & Grassini, 2020) and the ivory towers of digital nomad communities that lack real relationship development with local communities (Militello & Mirabile, 2023).

"SVIMEZ-Associazione per lo Sviluppo dell'Industria nel Mezzogiorno" ("Association for industry development in the South") is Southern Italy's main research center and has always focused, since its foundation in 1946, on the unfairness of the territorial divides (De Rita, 2020).

In 2020 SVIMEZ estimated that at least 40,000 workers returned to the South during the first year of the pandemic and at least 58,920 additional workers are potentially interested in the phenomenon (SVIMEZ, 2020). In particular, the survey conducted for SVIMEZ by Datamining measures 45,000 workers of large companies in the North who have been working remotely from the South since the beginning of the pandemic. A survey was carried out on a sample of 150 large companies, with over 250 employees, operating in various areas of the North in the manufacturing and service sectors. Considering also small and medium-sized enterprises (in Italy that means companies with more than 10 employees), it is estimated that the interested highly qualified workers may be as many as 100,000 (SVIMEZ, 2020).

The SW-LdS non-profit organization and movement, launched in March 2020, has a three-folded goal: stimulating policy proposals to make the phenomenon instrumental to the development of territorial cohesion; studying the phenomenon in an interdisciplinary way, through a Research Center (Osservatorio); and developing a "South Working Network" between stakeholders, i.e. workers, companies, and public institutions (Mirabile & Militello, 2022).

There are three infrastructural prerequisites identified by South Working for a municipality or a local entity to be able to join the "South Working Network".

The first prerequisite concerns the network connectivity infrastructure, linked to overcoming the so-called digital divide: a stable Internet connection of at least 20 Mbps (megabits per second) in download per single user is required, so that workers can communicate with other members of their workgroup via video calls, screen sharing, and collaborative whiteboards or documents in the cloud.

Secondly, at least a minimum transportation infrastructure is needed: a connection within 2 h from a main transportation hub (such as an airport or a station with high-speed trains), to enable workers to reach the employer's or customers' premises within a reasonable time in case of need.

Thirdly, the need for a social infrastructure was identified, i.e. a place from which one can work in a suitable environment, called a "community hub": a shared workspace from which to work while respecting social needs, in which to meet other remote workers but also local communities.

Among the most successful policy proposals is undoubtedly the creation of a network of these "community hubs". The main result of this initiative has been the activation of a network of spaces with workstations equipped for remote working (see below paragraph 5). These spaces are imagined not as simple workplaces not different from the traditional office but as real hubs for (new and pre-existing) local communities. They are places of participation, collaboration, innovation, and dialogue for workers, students, entrepreneurs, and local administrators, of education and play for children, of reading and meeting for the elderly (Gooch, 2022). They are not a "one-size-fits-all" model but are developed through horizontal governance leadership as a place-based policy (Barca et al., 2012). The goal is to make the communities whose fragility was worsened by COVID-19 stronger, more resilient, and more generative. In fact, having a designated place for participation, work and meetings can be a fundamental tool not only for local ecosystems but also for future "south workers" who will be able to adapt more easily to the existing context in the destination territories and to play a role of social activation and community involvement (Mariotti & Di Matteo, 2022).

A remarkable example of the "community hub" concept is located in the small mountain municipality of Castelbuono, 8000 inhabitants, 1 h drive from Palermo, the capital city of Sicily. There, a group of local South Workers convinced the local administration to create coworking spaces with 20 workstations in three spaces belonging to public heritage (two in museums and one in a building owned by the municipality in the main square of the town); negotiated with the TLC company Fiber Telecom a free WiFi connection (min. 20 Mbps); and created a dedicated website with a tool to reserve the workstations, as well as accommodations, restaurants, and other experiences (Fig. 5).

As to the perspectives on the future of the South Working phenomenon, it is necessary to highlight the main challenges faced, the limits of the current context, and the proposals for building a sustainable agile working model for companies, workers, and communities.

As part of the development of the network of community hubs, the main challenges faced in the last 5 years concern the relationship with local governments, the need to imagine a sustainable development of the territory with a long-term approach, the building of relationships of trust between stakeholders belonging to different sectors (non-profits, innovative companies, traditional businesses, local administrations, national legislators, etc.), and the maintenance of relationships among stakeholders over time. Adding to this already complex picture is an intricate national and international situation that undoubtedly affects local contexts, as in the case of inflation and the impact on the prices of goods and services. The generalized increase in prices affects Recovery Plan investment projects, as in the case of the

Fig. 5 One of the public community hubs in Castelbuono, Sicily (Chiostro San Francesco). Source: photo by the author

so-called "bando borghi" ("small towns grant"), i.e. approximately one billion euros for the cultural and social regeneration of small Italian towns. Among the main proposals that tend to stimulate the strengthening of local ecosystems through this grant is the creation of physical coworking spaces and the training of new professional profiles such as "community managers". They will need adequate skills to implement strategies, tools, and actions to facilitate relationships between the various communities, manage community hubs, activate processes of civic engagement and bottom-up participation, and develop a specific sensitivity oriented towards social, cultural, environmental, and economic sustainability.

Besides, the need for constant collaboration with and between institutions and other stakeholders emerges, reflecting the structural and infrastructural challenges of the Inner Areas. They concern the inefficient transportation from one municipality to another or from a small municipality to the closest city; other critical issues are linked to education and health services: for example, municipalities without a primary care provider and whose closest emergency room is located several kilometers away.

All these aspects are real obstacles to the implementation of the remote working and social support services offered by community hubs. The persistent lack of primary services can hold back innovative initiatives that can represent a guide for the improvement of territorial policies and welfare. A final need is to create integrated territorial development strategies between communities and institutions overcoming the traditional elaboration of "top-down" policies for the territories of Southern regions and Inner Areas from the central government. These are often followed by large resources and investment promises, which become a waste of public money because the municipalities themselves do not have the tools and resources they need to manage them.

Furthermore, there is a need to overcome the narrative of a hypothetical relationship of equivalence between remote working (from the South and internal areas) and tourism, instead concentrating social, political, and economic efforts to launch multi-stakeholder strategies to promote policies going beyond seasonality and favoring the so-called brain circulation against depopulation (Militello & Mirabile, 2023).

Finally, the creation of a regulatory framework and policies conducive to innovation and experimentation, e.g. adequate rules on remote working, represent a useful starting point for creating a more sustainable future for remote working, as happened elsewhere in the world (Bahar et al., 2022).

Methodology and Data Limitations

The present chapter uses data from three surveys conducted by the Italian non-profit "South Working-Lavorare dal Sud".

The first one was administered on an ongoing basis during the first and second phases of the pandemic to individuals potentially interested in working remotely from Southern regions, even after the health-related emergency. Participants were asked for information on work habits in the pre-Covid era and during it, together with other socio-demographic information, often obtained by asking questions on a 1–5 Likert scale.

Overall, the data frame contains 2594 observations on 93 variables. The time of collection ranges from June 2020 to December 2022. In particular, the survey obtained significantly more responses in the first months after its launch in June 2020 (Fig. 6).

The author presented the first batch of data analysis on this data frame in 2021, through an econometric *probit* analysis conducted through a two-step Heckman model, with significative results as to gender, family status, education level, and type of employer (public or private) but not on the type of employment (freelance or employee) (Di Matteo et al., 2021).

The second survey aimed at gathering data on existing coworking spaces and received 197 responses. Mostly, coworking owners participated in the survey, but also workers. The survey crowdsourced not only geolocation data but also the type of coworking spaces and the services offered for a total of 16 variables. The location of the space could be listed first as "urban area", "periurban area" or "inner area" and then, depending on the location within a municipality as "historic center", "urban center", "residential area", "periphery" or "midway between center and

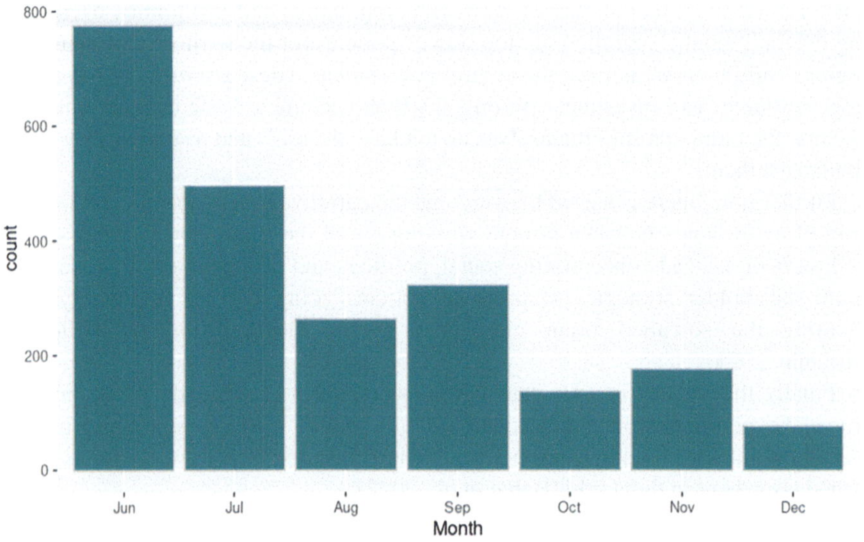

Fig. 6 Distribution of responses between June and December 2020. Source: Author

periphery". It then asked for the total number of admissible people in the space and the total number of workstations. The type of space could be specified, among "public" or "private", but also "big" or "small", depending on the number of workstations (more or less than 20), "shared space with single rooms", "workstation within a private company", "café", "restaurant", "coliving", "other private space" or "other public space (*e.g.* library)". Finally, they could add the available services (Wi-Fi, Optical fiber, shared desks, single-user desks, phone booths or areas for video calls, meeting rooms, library, lounge area, kids space, elderly space, changing table for infants and toddlers, nursing space, wheelchair accessibility, shared stationery, printer, projector, TV, shared chargers, shared tools, shared kitchen, outdoor areas, free drinking water, snacks, café, or others).

Additionally, 18 life stories were collected through an open-ended survey (maximum 2,200 characters) investigating the challenges young professionals faced while working remotely and their hopes for the future, asking how long they had lived away from their hometowns, when they moved back to the South or to their hometowns in general, how was their experience of South Working with its advantages and disadvantages, if they had perceived a lack of coworking spaces or had suffered from lack of services or inefficient transportation means. Due to the challenges of translation, the sentiment analysis and the word clouds will be tackled in a future research effort.

All data visualizations and processing were performed using the open-source programming language R (R Core Team, 2021), through the ggplot2 package (Wickham, 2016).

The main limitations of these three surveys concern the sample selection, as the distribution was based on a community organizing effort. This is a sample based on voluntary participation, shared through snowballing, national news outlets (D'Alessandro, 2020), and social media[1]. The data collected focuses mainly on bigger cities in the South of Italy, while it has been much more difficult to count and map smaller towns' return rates and coworking spaces (with remote working often improvised in cafés or other public spaces with an Internet connection) [2].

Preliminary Findings from a Comprehensive Set of Surveys on South Working in Italy

As to the first survey, the following main descriptive statistics of the variables used are presented through the following visualizations. Age was measured in age groups, and most of the respondents belonged to the 30–34 age group, closely followed by the 25–29 group and then the 35–39 one. More men than women filled out the survey. As to the family status, the vast majority of respondents were in a childless relationship. As to education, most of the respondents were highly educated, with most having a master's degree and several also having a second-level master's or a Ph.D. Finally, as to their work status, they were almost all workers and, among them, the majority had stable employment (Fig. 7).

In this first part of the analysis, a propensity index was defined and named "South Working propensity index" ("SWPI") (Fig. 8) summarizing three questions focused on the desire to perform "South Working":

"Did you want to come back to your home region before Covid?"—"wish_to_comeback";
"Did you want to move South before Covid?"—"wish_to_live_South";
"Would you move to the South if you could maintain your job and work remotely?"—"SW_propensity".

The resulting index can be expressed as follows:

$$SWPI_i = \frac{wish\_comeback_i + wish\_live\_south_i + SW\_propensity_i}{3}$$

and ranges from 1 (minimum propensity) to 5 (maximum propensity).

The results show a strongly skewed curve towards the highest propensity (Fig. 9).

---

[1] The "*South Working – Lavorare dal Sud*" Facebook page, which devised and distributed the surveys, has 13,319 followers, of which 48.5% are men and 51.5% are women, mainly in the age groups 25–34 and 35–44 and in the Southern towns of Palermo, Catania, Messina, Naples, Bari; and in the Northern/Central urban centers of Milan, Rome, Bologna, Turin (Data: Facebook Insights, 03.04.2023).

[2] Other for-profit organizations like *www.italiancoworking.it* developed wider observations, counting over 800 coworking spaces in Italy (albeit without checking whether their founding values had any social implication or "give back" to the community goals).

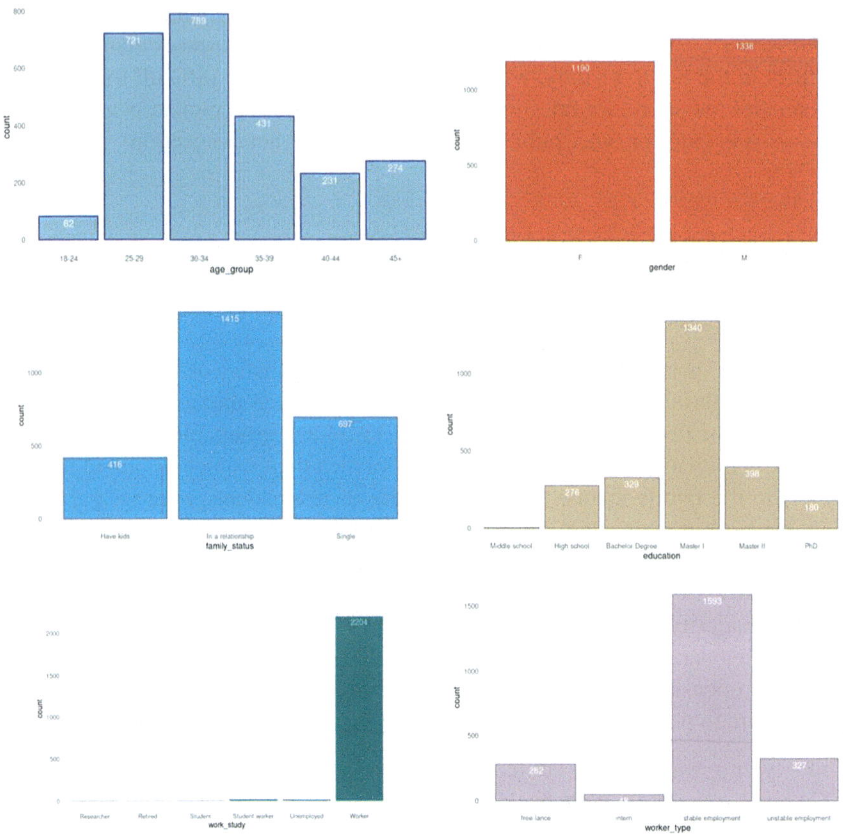

Fig. 7 Distribution of respondents by age, gender, family status, education, occupation and type of employment. Source: Author

We then elaborated on the distribution of the index SWPI in relation to the six socioeconomic variables highlighted at the beginning (Fig. 10).

As to gender, women with the highest propensity index score (5 out of 5) are 39% of the total of female respondents, while men with the same score are 41.7% of all male respondents.

Secondly, to answer the research question on the phenomenon's distribution in the Italian regions, South Working has deployed a crowdsourcing mapping technique to gather geolocation attributes of coworking hubs.

From the second survey, the one on coworking spaces or "community hubs", collected data was employed to create a participatory mapping of coworking spaces that is publicly available online. It highlights especially those coworking spaces adhering to the South Working Network of Community Hubs (the "Network"), showing services offered by individual coworking spaces sharing the value of giving back to the local communities (Fig. 11).

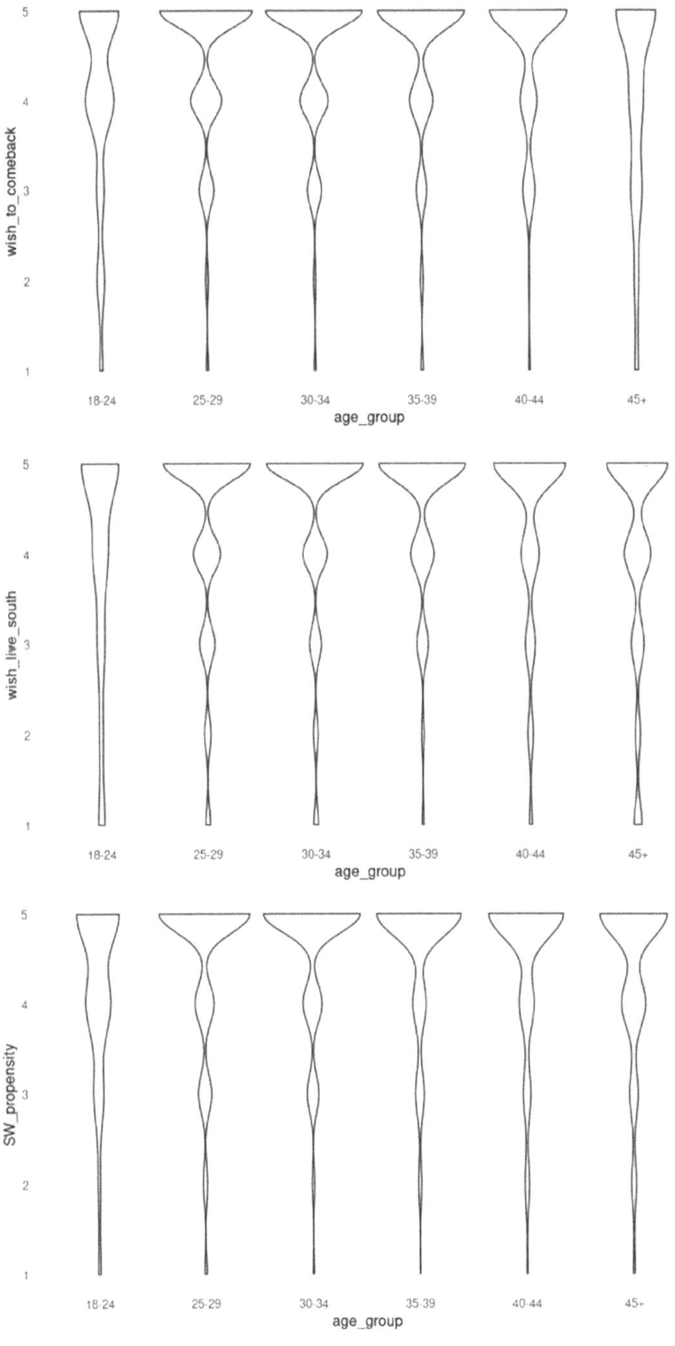

Fig. 8 Violin plot of distribution by age group of the three indicators. Source: Author

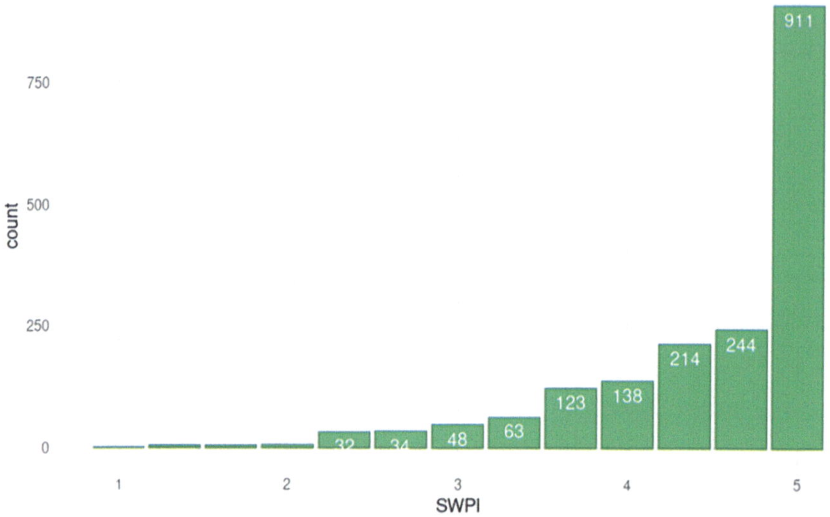

Fig. 9 South Working Propensity Index. Source: Author

There are 230 pinpointed coworking spaces, 11 of which are municipalities that participate in the Network and 38 are private coworking spaces members of the Network.

Among the different typologies, it is possible to find: coworking spaces, impact hubs, rural hubs, spaces for socializing, private companies with spaces open for coworkers, and 'new' public spaces, such as libraries, underused rooms of museums, and former convents. All these places are characterized by a natural aggregative, associative, creative, and innovative dimension. Community hubs have the purpose of establishing a social infrastructure useful for the creation of relational networks of people and organizations, local communities, workers, associations, administrations, companies, and cultural institutions, all committing to co-design activities and services for the sustainable development of the community, through the establishment of new businesses, support for the administration and citizens through trainings, networking, and dissemination of knowledge.

Collected data shows that only 9% of mapped coworking spaces are located in Inner Areas (according to their users and owners as it is a piece of self-reported information).

Conclusions and Policy Proposals

The present chapter, informed by extensive reflections and analyses, sought to identify who seems more predisposed to experience remote working, delineate the feasible locations for remote working, and explore how public and private policies can

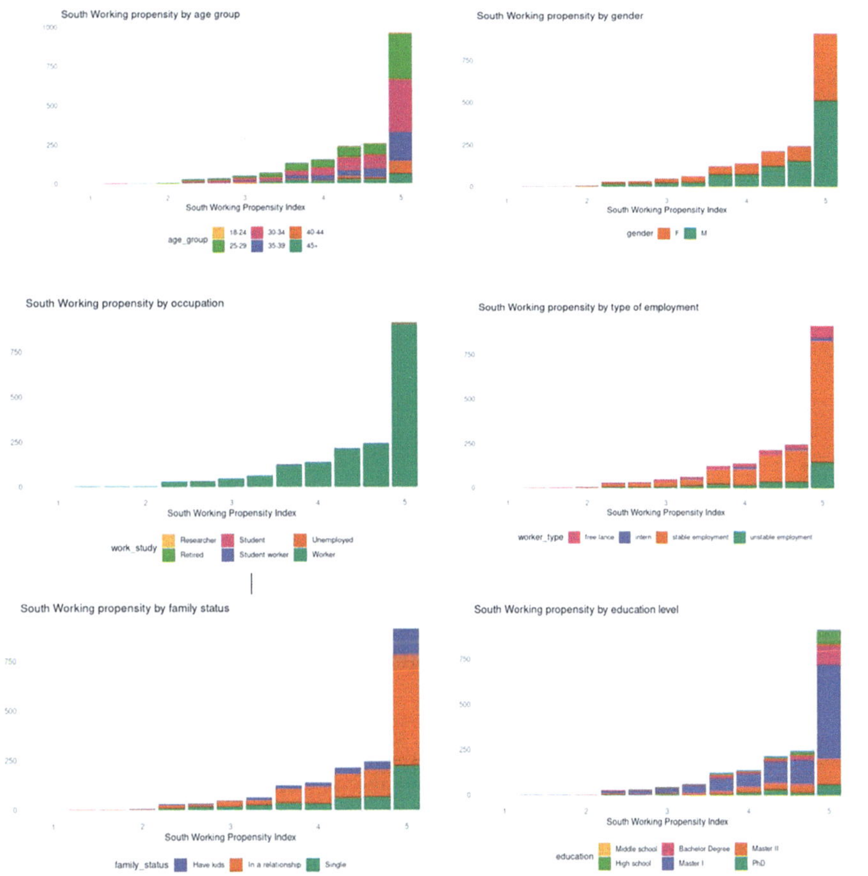

Fig. 10 South Working Propensity Index by age group, gender, family status, education, occupation, and type of employment. Source: Author

effectively funnel the available financial and human capital resources to support a sustainable form of remote working.

This investigation broadens the current debate on potential new public policies to implement and enhance the experience of workers seeking to engage in “South Working” periods in different global regions and, specifically, in Italy.

Mitigation of the digital divide and strengthening of digital infrastructures represent primary goals of the “Southern Plan 2030” presented in February 2020 by the Italian Minister for South and Territorial Cohesion.[3] The Recovery Plan itself has

[3] Italian Agency for Territorial Cohesion (“*Agenzia per la coesione territoriale*”), *Piano Sud 2030. Sviluppo e coesione per l’Italia*, 6 March 2020, https://www.agenziacoesione.gov.it/download/piano-sud-2030-sviluppo-e-coesione-per-litalia/.

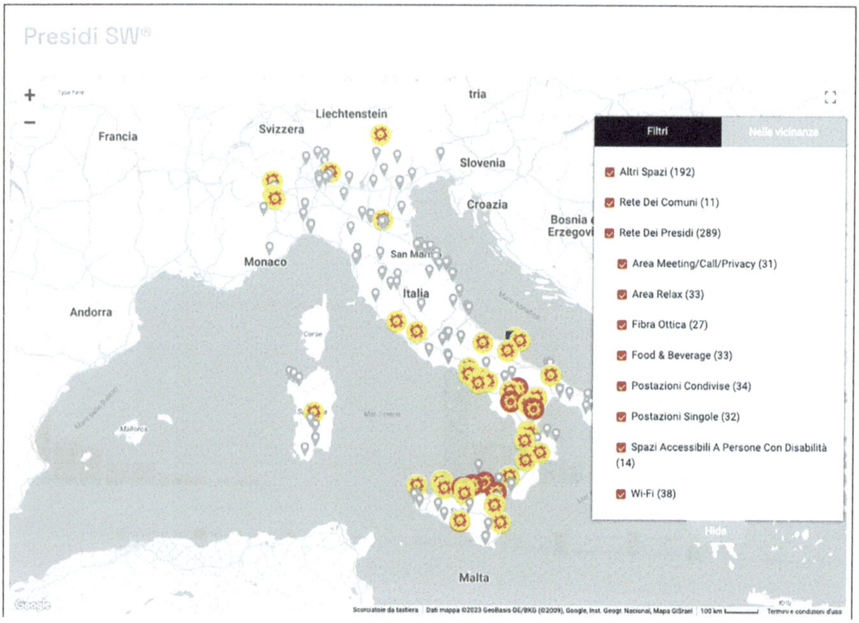

Fig. 11 South Working Network of Community Hubs. Source: data and cartography by South Working - Lavorare dal Sud A.P.S

made a similar commitment to these strategic infrastructures, as evidenced by remote learning and working during and after the pandemic.

The present analysis suggests that the phenomenon of "South Working", if strategically placed within a policy framework prioritizing local contextual conditions (from material to social infrastructures, in close conjunction with providing public services), could be a significant opportunity to improve the currently existing territorial gaps and counter the de-accumulation of qualified human capital witnessed over more than 20 years in the South and in the Inner Areas of the country.

Furthermore, this analysis shows that remote working cannot be imposed simply by relegating workers to their own homes, often in inadequate spaces and untenable family conditions. Instead, it requires policies for the creation of coworking spaces and family support. Remote workers need innovative workplaces that reduce the sense of isolation often associated with remote, while also offering support services (*e.g.* wi-fi connection, well-equipped meeting rooms, communal kitchen facilities and leisure spaces, access to training and coaching programs, and childcare provisions like nurseries or shared nannies). Such provisions are crucial for balancing personal and work life (Manzini Ceinar & Mariotti, 2021; Mariotti et al., 2022a).

This suggestion also serves to mitigate potential and unintended negative consequences on living costs, like gentrification. Activating coworking spaces as intermediaries between remote workers and local communities might incentivize a shared motivation for "giving back" and avoid the exploitation of regions with a lower cost of living.

Ultimately, the integration of "South Workers" and the development of co-working spaces can foster dynamic local ecosystems, facilitating a productive exchange between externally acquired skills and locally grounded experiences.

The peculiarity of the suggested "South Working" proposal, compared to pre-existing phenomena such as digital nomadism or traditional teleworking, lies in its explicit mission, aimed at improving economic, social, and territorial cohesion. The influx of human (highly skilled) capital into marginalized economies would offer an immediate benefit through increased local liquidity, driven by the greater demand for goods and services and from the greater purchasing power and consumption capacity of these individuals. Furthermore, the permanence of these subjects in these areas could bring long-term benefits in terms of investments and democratic participation, including advocating and pressuring local institutions to better manage resources and provide more efficient services. Ultimately, the presence of South Workers and coworking spaces is set to lead to the creation of stimulating ecosystems in local communities, in a fertile dialogue between externally acquired skills and locally grounded experiences.

Acknowledgements I owe a great debt of gratitude to the Minda de Gunzburg Center of European Studies at Harvard University, where I conducted this research as a Visiting Scholar. I am particularly thankful to the cohort of colleagues who participated in my New Research on Europe Seminar on April 3, 2023, and provided fundamental feedback on this chapter. I extend my sincere thanks to Professor Prithwiraj Choudhury of the Harvard Business School for kindly agreeing to chair the seminar. In a parallel way, I express my profound gratitude to every South Worker and every associate of the "South Working" nonprofit organization and movement, which has provided me with a renewed purpose over the last 5 years, and with whom we gathered all the data presented, and especially the Vice-President, Mario Mirabile, and my right-hand woman, Flavia Amoroso.

References

Angelici M., & Profeta P. (2020). *Smart-working: Work flexibility without constraints*. CESifo Working Paper No. 8165.

Badia, A., Langemeyer, J., Codina, X., et al. (2021). A take-home message from COVID-19 on urban air pollution reduction through mobility limitations and teleworking. *Urban Sustainability, 1*(35).

Bahar D., Choudhury P., Signorelli S., and Sappenfield J. M. (2022) Talent Flows and the Geography of Knowledge Production: Causal Evidence from Multinational Firms. HBS Working Paper 22047.

Balabio, B. (2022). Gli smart worker italiani scendono a 3,6 milioni, previsioni di crescita nel 2023. Osservatori.net, 20.10.2022, www.osservatori.net. Last accessed 2025/01/03.

Barbieri, T., Basso, G., & Scicchitano, S. (2020) *Italian workers at risk during the Covid-19 epidemic*. INAPP working paper, n. 46.

Barca, F., McCann, P., & Rodríguez-Pose, A. (2012). The case for regional development intervention: Place-based versus place-neutral approaches. *Journal of Regional Science, 52*, 134–152.

Bellandi, M., Mariotti, I., & Nisticò, R. (Eds.). (2021). *Città nel Covid. Centri urbani, periferie e territori alle prese con la pandemia*. Donzelli.

Bianchi, L., & Fraschilla, A. (2020). *Divario di cittadinanza*. Un viaggio nella nuova questione territoriale.
Bianchi, L., Petraglia, C., & Vecchione, G. (2019). Europa-Italia, Nord-Sud: il doppio divario. *Rivista economica del Mezzogiorno, 3-4*, 673–687.
Bianchi, L., & Vecchione, G. (2021). Vivo al Sud, lavoro al Nord. In M. Bellandi, I. Mariotti, & R. Nisticò (Eds.), *Città nel Covid. Centri urbani, periferie e territori alle prese con la pandemia*. Donzelli.
Botsari, K. (2022). *The new legal framework for teleworking in Greece*. Mondaq. www.mondaq.com. Last accessed 03 Jan 2025
Bubola, E. (2021, January 8). The pandemic helped reverse Italy's brain drain. But can it last?. *The New York Times*. www.nytimes.com. Last accessed 03 Jan 2025.
Cascaldi-Garcia D., Orak M., Saijid Z. (2023, January 13) Drivers of post-pandemic Inflation. In: *Selected advanced economies and implications for the outlook, FEDS Notes*. www.federalreserve.gov. Last accessed 2025/01/03.
Casti, E., & Adobati, F. (Eds.). (2021). *Mapping the epidemic: A systemic geography of COVID-19 in Italy*. Elsevier-Health Sciences Division.
Cersosimo, D., & Donzelli, C. (2020). *Manifesto per riabitare l'Italia*. Donzelli.
Chiellino, G. (2023). Neet, alle regioni italiane il record europeo. *Il Sole 24 Ore*. lab24.ilsole24ore.com. Last accessed 03 Jan 2025
Choudhury, P. (2022a). Tulsa remote: Moving talent to middle America. *Harvard Business School Teaching Note*, 623–638.
Choudhury, P. (2022b). Geographic mobility, immobility, and geographic flexibility—A review and agenda for research on the changing geography of work. *Academy of Management Annals, 16*(1), 258–296.
Choudhury P., Kwon O. (2018) Social Attachment to Place and Psychic Costs of Geographic Mobility: How Distance from Hometown and Vacation Flexibility Affect Job Performance. HBS Working Paper 19010.
Choudhury, P. (2020). Our work-from-anywhere future. Best practices for all-remote organizations. *Harvard Business Review Magazine*. November–December 2020, hbr.org. Last accessed 03 Jan 2025.
Choudhury, P., Foroughi, C., & Larson, B. (2021). Work-from-anywhere: The productivity effects of geographic flexibility. *Strategic Management Journal, 42*(4), 655–683.
Choudhury P., Sevcenko V., Khanna T. (2018) Should Firms Move Talent from the Geographic Periphery to Hubs? A Strategic Human Capital Perspective. HBS Working Paper 14–080.
Corte dei Conti. (2021, May). *Referto sul Sistema Universitario 2021*. www.aranagenzia.it. Last accessed 03 Jan 2025.
Corso M. (2020). Sono otto milioni. Ma la vera sfida comincia adesso. *la Repubblica*. 25 maggio 2020. https://www.repubblica.it/dossier/tecnologia/rivoluzione-smartworking/2020/05/25/news/otto_milioni_di_lavoratori_in_lavoro_agile_ma_la_vera_sfida_comincia_adesso-257510754/
Cracolici, M. F., & Nijkamp, P. (2009). The attractiveness and competitiveness of tourist destinations: A study of Southern Italian regions. *Tourism Management, 30*(3), 336–344.
Croce, G., & Scicchitano, S. (2022). Cities and working from home in Italy in the post COVID-19 age. *Rivista Italiana di Politiche Pubbliche, 17*(2), 191–217.
D'Alessandro, J. (2020). Dallo smart working al south working. "Per lavorare a Milano vivendo a Palermo". *Repubblica*. www.repubblica.it. Last accessed 03 Jan 2025
De Rita, G. (2020). Il lungo Mezzogiorno. Interpretazioni e narrazione. Antologia 1966–2002. Laterza ed.
Di Matteo, D., La Regina, R., Mariotti, I., & Militello, E. (2021). Quali determinanti per il South Working? Una nuova proposta di sviluppo per il Sud, le aree interne e il Paese. *Rivista economica del Mezzogiorno, 4*, 678–701.
Dotti, N. F., Fratesi, U., Lenzi, C., & Percoco, M. (2013). Local labour markets and the interregional mobility of Italian University students. *Spatial Economic Analysis, 8*, 443–468.

Elton, C. Portugal welcomes hundreds of digital nomads under new visa scheme. Euronews. 26/01/2023. https://www.euronews.com/travel/2023/01/26/digital-nomads-will-be-able-to-live-and-work-in-portugal-full-timewith-this-new-visa
Encarnación, O. G. (2022). The revolt of empty Spain. *NY Review of Books*. www.nybooks.com. Last accessed 03 Jan 2025
Eurofound-ILO (2017) Working anytime, anywhere: The effects on the world of work, Publications Office of the European Union, Luxembourg, and the International Labour Office, Geneva.
Eurofound (2021) 2020 European Working Conditions Survey (EWCS), www.eurofound.europa.eu, last accessed 2025/01/03.
EUROSTAT (2019) European Union Labour Force Survey,. ec.europa.eu. Last accessed 03 Jan 2025.
Fana M., Milasi S., Napierala J. et al. (2020) Telework, work organisation and job quality during the COVID-19 crisis: a qualitative study, JRC Working Papers Series on Labour, Education and Technology No. 2020/11, European Commission, Joint Research Centre (JRC), www.econstor.eu, last accessed 2025/01/03.
Felstead, A., & Henseke, G. (2017). Assessing the growth of remote working and its consequences for effort, well-being and work-life balance. *New Technology, Work, and Employment, 32*(3), 195–212.
Florida, R. L. (2002). *The rise of the creative class: And how it's transforming work, leisure, community, and everyday life* (1st ed.). Basic Books.
Florida, R. L., & Ozimek, A. (2021). How remote work is reshaping America's urban geography. *Wall Street Journal*. www.wsj.com. Last accessed 03 Jan 2025
Frost, R. (2023). Spain's digital nomad visa is finally available. Here's everything you need to know. *Euronews*. www.euronews.com. Last accessed 03 Jan 2025
Ganga, R., et al. (2018). From Portugal to Europe. A micro-level sociology of scientific migration in times of Eurozone crisis. *Migraciones Internacionales, 9*(4), 9–37.
Ghiglione D., Romei V. (2021) Italian returnees seize on pandemic to stop Mezzogiorno brain drain. Financial Times, 6 April 2021. www.ft.com, last accessed 2025/01/03.
Glaeser, E. (2011). *The triumph of the city*. Palgrave Macmillan.
Giambona, F., & Grassini, L. (2020). Tourism attractiveness in Italy: Regional empirical evidence using a pairwise comparisons modelling approach. *International Journal of Tourism Research, 22*, 26–41.
Giuzio W., & Rizzica L. (2021, January 22). *Il lavoro da remoto in Italia durante la pandemia: le amministrazioni pubbliche, Note Covid-19 Banca d'Italia*. www.bancaditalia.it. Last accessed 03 Jan 2025.
Gooch, A. (2022). The new geography of work: Realising the promises of remote work for local development. *OECD-Forum*. www.oecd-forum.org. Last accessed 03 Jan 2025
Gramsci, A. (1991). *La questione meridionale*. Editori Riuniti, or. 1926.
ISTAT (2020) Rapporto annuale 2020, .
Junker, T. L., Bakker, A. B., Gorgievski, M. J., & Derks, D. (2022). Agile work practices and employee proactivity: A multilevel study. *Human Relations, 75*(12), 2189–2217.
Kaysen, R. (2023). Americans head to Europe for the good life on the cheap. *New York Times*. www.nytimes.com. Last accessed 03 Jan 2025
Leitão, M. (2022). Portugal: New rules on remote work in Portugal effective January 2022. *L&E Global*. leglobal.law. Last accessed 03 Jan 2025
Leitner, S. (2021). Evidence of brain gain for some Western Balkan countries. *The Vienna Institute for International Economic Studies*. wiiw.ac.at. Last accessed 03 Jan 2025
Lucatelli, S., Monaco, F., & Tantillo, F. (2019). La Strategia delle aree interne al servizio di un nuovo modello di sviluppo locale per l'Italia. *Rivista economica del Mezzogiorno, 3-4*, 741 ff.
Manzini Ceinar, I., & Mariotti, I. (2021). Teleworking in post-pandemic times: May local coworking spaces be the future trend? *Romanian Journal of Regional Science, 15*(1), 52–76.
Mariotti, I., Akhavan, M., & Rossi, F. (2023). The preferred location of coworking spaces in Italy: An empirical investigation in urban and peripheral areas. *European Planning Studies. 31*(3), 467–489.

Mariotti, I., & Di Matteo, D. (2022). Are coworkers in the Italian peripheral areas performing better? A counterfactual analysis. *Sustainability, 14*(1), 550 ff.
Mariotti, I., Di Matteo, D., & Rossi, F. (2022a). *Who loses and who gains during the Covid-19 pandemic? The rise of remote working in suburban areas*. Mimeo.
Mariotti, I., Di Marino, M., & Bednar, P. (Eds.). (2022b). *The COVID-19 pandemic and future of working spaces*. Routledge.
Mariotti, I., Giavarini, V., Rossi, F., & Akhavan, M. (2022c). Exploring the "15-minute city" and near working in Milan using mobile phone data. *TeMA—Journal of Land Use, Mobility and Environment*, 39–56.
Militello, E., & Mirabile, M. (2020). Focus - Il progetto «South working» – Lavorare dal Sud. *Rapporto SVIMEZ—L'economia e la società del Mezzogiorno. il Mulino.*, 223–234.
Militello, E., & Mirabile, M. (2023). Smart working, the south and the inner areas for local development. *E & M, 1*. emplus.egeaonline.it. Last accessed 03 Jan 2025
Mirabile, M., & Militello, E. (2022). South working: per un futuro sostenibile del lavoro agile in Italia. *Donzelli*.
Mohammad, L., Jarenwattananon, P., & Summers, J. (2023). An open letter signed by tech leaders, researchers proposes delaying AI development. *NPR*. www.npr.org. Last accessed 03 Jan 2025
Moretti, E. (2012). *The new geography of jobs* (1st ed.). Houghton Mifflin Harcourt.
Murillo Chávarro, J. (2023). Work-from-home regulations are coming. Companies aren't ready. *MIT Sloan Management Review*. sloanreview.mit.edu. Last accessed 03 Jan 2025
Musolino, D., Meester, W. J., & Pellenbarg, P. H. (2021). Stated locational preferences of Italian entrepreneurs: The underlying location factors. *Regional Science. Policy & Practice*, 1–17.
Musolino, D., & Rizzi, P. (2020). COVID-19 e territorio: un'analisi a scala provinciale. *EyesReg, 10*(3).
Neeley, T. (2021). *Remote work revolution: Succeeding from anywhere* (1st ed.). Harper Business.
Nelson, O. (2015). The social effects of the Spanish brain drain. *Social Impact Research Experience (SIRE)., 35*.
Nifo, A., & Vecchione, G. (2014). Do institutions play a role in skilled migration ? The case of Italy. *Regional Studies, 48*(10).
Nilles, J. M., Carlson, F. R., Gray, P., & Hanneman, G. J. (1976). *The telecommunications-transportation tradeoff: Options for tomorrow*. Wiley.
OECD. (2016). *Be flexible. Background brief on how workplace flexibility can help European employees to balance work and family*. www.oecd.org. Last accessed 03 Jan 2025.
Openpolis-Con i Bambini. (2019). Le grandi regioni del sud hanno più abbandoni e più neet. www.openpolis.it. Last accessed 03 Jan 2025.
Osservatorio Smart Working del Politecnico di Milano. (2019, October 30). *Cresce il numero degli smart worker italiani: 570mila (+20%) lavoratori più soddisfatti degli altri*. www.osservatori.net. Last accessed 03 Jan 2025.
Osservatorio Smart Working del Politecnico di Milano (2020, November 3). *Dallo smart working d'emergenza al "New normal": nuove abitudini e nuovi approcci al lavoro*. www.osservatori.net. Last accessed 03 Jan 2025.
Padilla, K. J. (2022). *An Analysis of the Strengths and Weaknesses of the Tulsa Remote Program, as an Effective Economic Development Strategy*. ProQuest Dissertations Publishing.
Pimpinella A., Di Giusto F., Redondi A., Venturini L., Pavon A. (2022) Forecasting busy-hour downlink traffic in cellular networks. Working paper.
Pratsinakis, M. (2022). Greece's Brain Drain Problem Isn't Going Away. *World Politics Review*. www.worldpoliticsreview.com. Last accessed 03 Jan 2025
PWC-Italy Research Center (2021). *Smart working. Position paper*. PWC-Italy.
R Core Team. (2021) R: A language and environment for statistical computing. R foundation for statistical computing. https://www.R-project.org/. Last accessed 03 Jan 2025.
Recchi, E., & Triandafyllidou, A. (2010). Crossing over, heading west and south: Mobility, citizenship, and employment in the enlarged Europe. In G. Menz (Ed.), *Labour migration in Europe*. Springer.

Recchi, E., Barone, C., & Assirelli, G. (2016). Graduate migration out of Italy: Predictors and pay-offs. *Notes & Documents de l'OSC, n. 2016-03.*

Rodriguez, C. (2022). Live and work in Europe: More countries, including Portugal, offering trendy digital nomad visas. *Forbes*. www.forbes.com. Last accessed 03 Jan 2025

Salvemini, G. (1955). *Scritti sulla questione meridionale 1896–1955*. Einaudi.

Sanfilippo, M. (2015). Nuovi studi sulle migrazioni interne. *Studi Emigrazione, 199*, 483–488.

Schneider J. (ed.) (1998) Italy's 'Southern Question'. Orientalism in One Country, Routledge, 1st ed.

Sisson, P. (2019). Reversing the rural brain drain with remote working. *Curbed*. archive.curbed.com. Last accessed 03 Jan 2025

Sonzogno, G. V., Urso, G., & Faggian, A. (2022). Migration propensity of peripheral youth: Insights from Italy. *Regional Studies, Regional Science, 9*(1), 709–726.

Sostero, M., Milasi, S., Hurley, J., Fernandez-Marcias, E., & Bisello, M. (2020). *Teleworkability and the COVID-19 crisis: A new digital divide?* Sevilla, EU Commission, JRC, Eurofound.

Speciale, A. (2023, March 27). *EU Delays Disbursement of €19 Billion of Pandemic Aid to Italy*. Bloomberg. www.bloomberg.com. Last accessed 03 Jan 2025.

SVIMEZ. (2019). *Rapporto SVIMEZ 2019. L'economia e la società nel Mezzogiorno*. Il Mulino.

SVIMEZ. (2020). *Rapporto SVIMEZ 2020. L'economia e la società nel Mezzogiorno*. Il Mulino.

SVIMEZ. (2022). *Rapporto SVIMEZ 2022. L'economia e la società nel Mezzogiorno*. Il Mulino.

Treccani. (2020). South working. In *Neologismi*. www.treccani.it. Last accessed 03 Jan 2025

Vecchione, G. (2017). Migrazioni intellettuali ed effetti economici sul Mezzogiorno d'Italia. *Rivista economica del Mezzogiorno, 3*, 643–662.

Wickham, H. (2016). *ggplot2: Elegant graphics for data analysis*. Springer.

Digital Valtellina: A Regional Plan to Revitalize Italy's Inner Areas Through Remote Work

Surbhi Agrawal and Elaine Wang

Introduction

The Valtellina region, a valley in the foothills of the Italian Alps, is facing serious challenges of economic degrowth and population decline. In this project, we explore how the global trend of remote work, which has been accelerated by the COVID-19 pandemic, can be used as a catalyst for economic revitalization in this area. How might we revitalize Valtellina's local economies and develop tourism through remote work?

This project proposes a regional plan to revitalize Valtellina's local economies and develop tourism through remote work. To inform this plan, we conducted workshops with remote workers to understand their needs. We then engaged with the region's municipal authorities to learn about Sondrio's existing plans for a "Smart City" and discover the region's potential to develop as a destination for remote work. Finally, we developed a city-town-village development strategy, a place-based framework that proposes physical, digital, and policy interventions for the various scales of towns in the region (Fig. 1). The key principles of our strategy are to develop a "Remote Work Valley", build a new knowledge economy, and further strengthen sustainable tourism in the region.

S. Agrawal
Sasaki USA, Boston, MA, USA

MIT Masters of City Planning 2023, Cambridge, MA, USA
e-mail: surbhiag@mit.edu

E. Wang (✉)
Jane Energy, New York, NY, USA

MIT Masters of City Planning 2023, Cambridge, MA, USA
e-mail: elaine@jane.energy

G. Fossa et al. (eds.), *Small-Town Renaissance*, Research for Development,
https://doi.org/10.1007/978-3-031-89786-3_7

Fig. 1 Vision for Digital Valtellina. Source: image by the authors

Context

Valtellina is a valley in the Lombardy region of northern Italy, bordered by the Bernina Alps, which is part of the Rhaetian Alps. It spans a 132-miles east–west axis from the village of Berbenno to the town of Tirano in the Sondrio district and forms a part of the border between Italy and Switzerland (Micallef, 2018). It is close to Milan, Italy's capital city, which is easily accessible by both road and rail (Fig. 2).

Valtellina has a stunning natural landscape of mountains and valleys, which allows for year-round activities like skiing in the winter and biking in the summer. It also has a rich history and cultural heritage, with vineyards, historic architecture, and culinary offerings throughout (Fig. 3).

However, Valtellina faces several challenges. Its local economies are shrinking due to population decline and changing industries. Most cities and towns in the province are faced with population degrowth between 0.5% and 3% (Fig. 4). Traditional industries like manufacturing, construction, trade and agriculture have declined over the past decade with some growth in service and tourism-based industries (infocamere.it, 2022).

There is a lack of jobs in Valtellina, and most young people move to larger cities like Milan. Most cities and towns in the region have low average incomes, low concentrations of workers, few local enterprises, and low tourism density (Fig. 5). Most economic activity is concentrated in the province's capital city of Sondrio, as well as skiing towns such as Livigno, Bormio and Chiavenna. The region also lacks high-speed digital connectivity. The Federal Communications Commission (FCC)

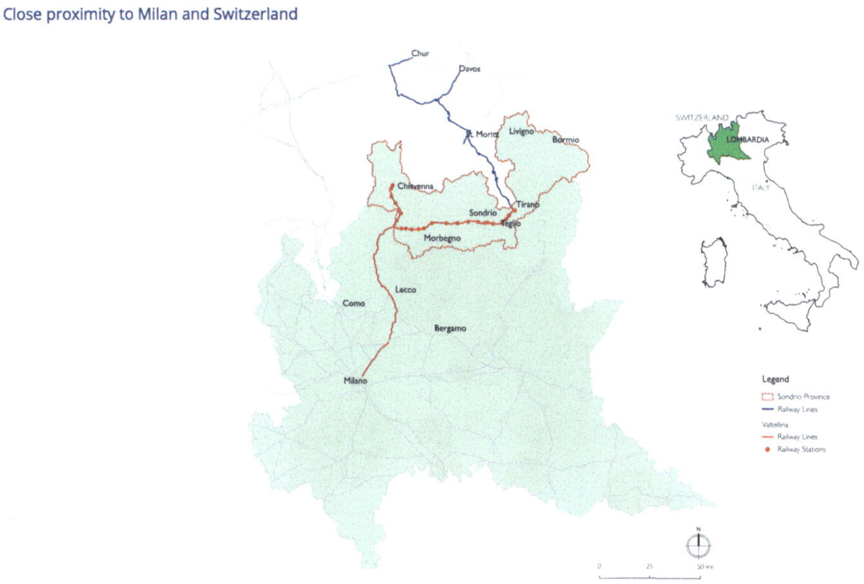

Fig. 2 Location of Valtellina. Source: image by the authors

Fig. 3 Valtellina's cultural and culinary offerings. Source: image by the authors

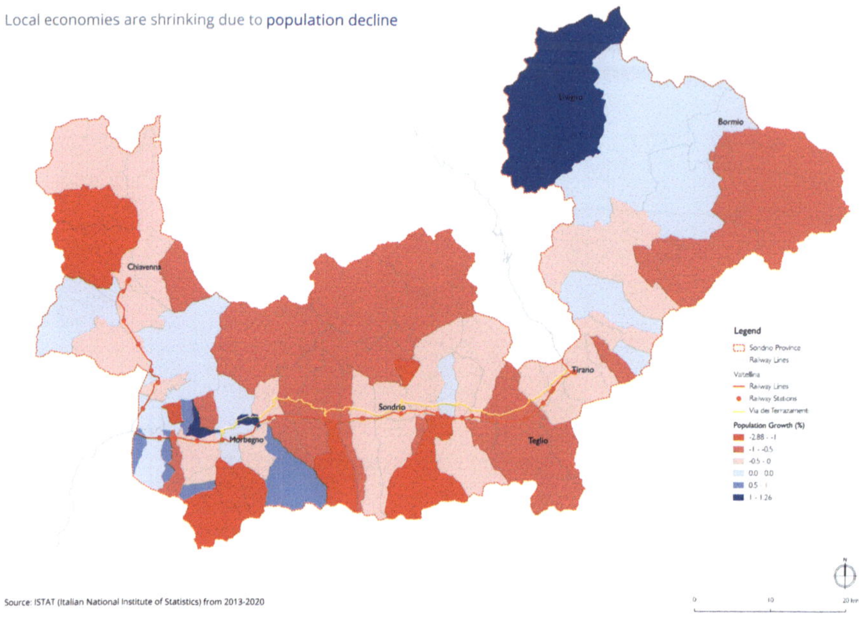

Fig. 4 Population decline in Sondrio Province between 2013 and 2020. Source: image by the authors

defines broadband internet as a minimum of 25 Mbps download speed and 3 Mbps upload speed ("Internet Speed Classifications | CWS Blog," 2020). At present, Sondrio does not have the high-speed broadband internet infrastructure required to support remote work and digitally supported economic activity (Fig. 6).

On the tourism front, there is a lot of competition with nearby destinations as the seasonal activities are very similar. Nearby Alpine regions in Switzerland and Austria attract most of the tourism in the Alps. Despite efforts by the regional authorities to promote the region as a tourist destination for cultural experiences, natural landscapes, and outdoor activities like skiing, hiking, and cycling, Valtellina has an incoherent and disconnected regional experience and lacks a distinctive attractive identity.

Remote Work as a Solution

Is remote work an opportunity for Italy's inner areas? During the COVID-19 pandemic, many small cities and towns in Italy experienced a rise in remote work, driven by young workers returning to their hometowns during the lockdown. Many people in Milan moved out to smaller cities and to second homes in the Lombardy region. This shift is evident in the development of working groups, co-working hubs and co-living spaces throughout Italy and neighboring countries. To learn more

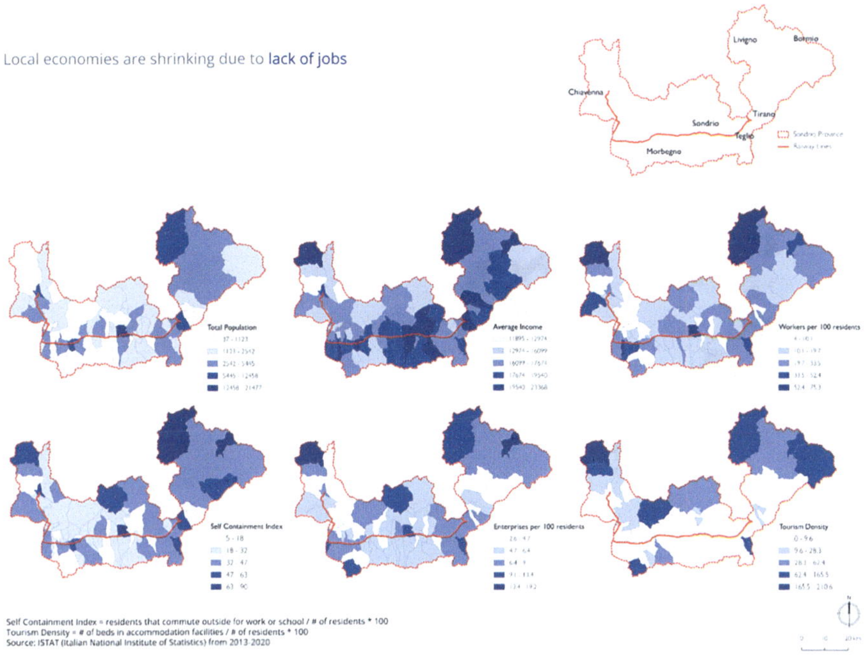

Fig. 5 Shrinkage of the region's local economies. Source: image by the authors

about remote work trends in Italy, we spoke with Elena Militello, the president of South Working, an Italian non-profit association that was founded in 2020. South Working is working on strengthening the local economy of the Southern regions of Italy by leveraging remote work opportunities. Through our project, we consider the opportunity for socio-economic development in Valtellina through the rise of the remote work.

Valtellina is already exploring the potential of remote work. In our conversations with the members of Sondrio's city council (Fig. 7), we learned that Sondrio is experimenting with hosting larger events and co-working space. The vision for the city is to become a gateway to smaller towns in the region and a "Smart City". The council is also planning to leverage the upcoming Winter Olympics to boost seasonal tourism and winter sports activities in the region, including building flexible housing options, improving digital connectivity, and supporting local entrepreneurship.

To better understand this opportunity, we engaged with remote workers to learn more about what they look for in a destination. We sent out a survey in November 2022 through social media to remote workers in Italy and abroad. Our survey focused on questions around what remote workers look for in a destination, including factors like community engagement, type of place, internet access, ease of travel, living costs, proximity to nature, etc. We received 60 responses in total. From our survey, we found that (Fig. 8):

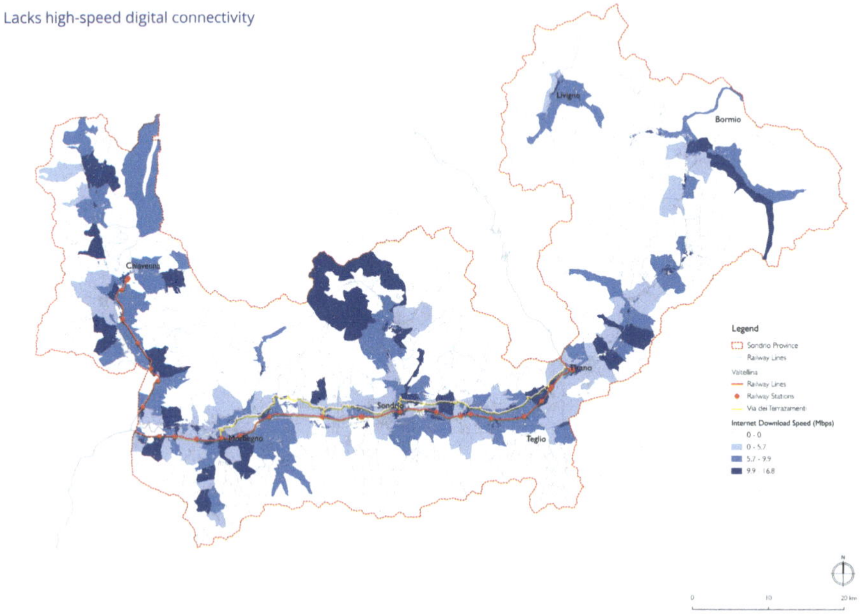

Fig. 6 Internet Download Speeds in Sondrio (2022). Source: image by the authors; data: AGCOM Broadband Map

Fig. 7 Our conversation with Sondrio's city council. Source: image by the authors

1. Most respondents were fully remote or hybrid.
2. Most respondents have worked remotely from abroad.
3. Most respondents wanted to work remotely for a few months.
4. Most were interested in local engagement.
5. There was lesser interest in places "off the beaten path".
6. There was a lot of interest in community engagement.
7. Internet, ease of travel, and costs are the most important factors for remote work.

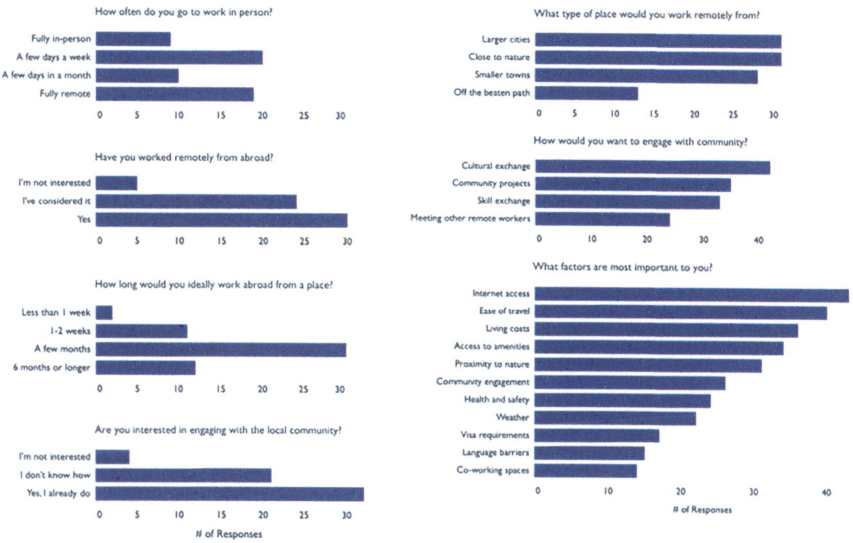

Fig. 8 Results from online survey for remote workers. Source: image by the authors

Fig. 9 Online workshop with remote workers. Source: image by the authors

8. Factors like amenities and nature, and community engagement were also important to the respondents.

We also conducted an online workshop with digital nomads in Italy (Fig. 9) that we found through Facebook groups. Through the workshop, we got insight into what the digital nomad experience is like, why they chose Italy, what their problems are, and what they look for in a place.

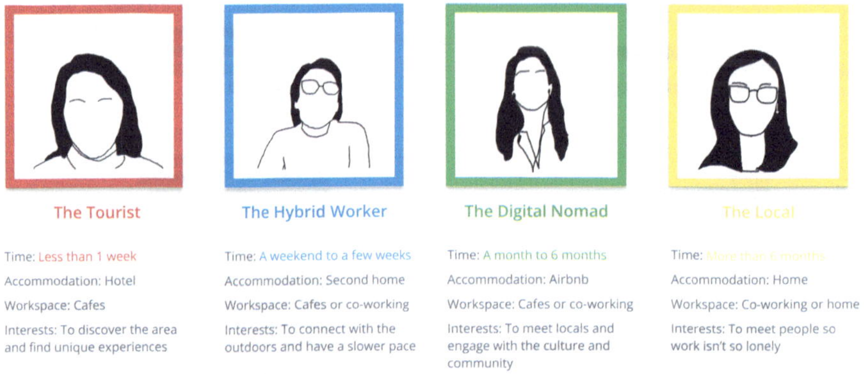

Fig. 10 Types of remote workers. Source: image by the authors

This workshop uncovered some new findings. We found that remote workers want high-speed internet, local activities and amenities, affordable cost of living, flexible housing options, and low taxes. People also shared that remote work can sometimes be isolating and a lot of effort is required to engage with locals. Based on these discussions, we learned that there are many types of remote workers (Fig. 10). We categorized them as:

1. The Tourist.
2. The Hybrid Worker.
3. The Digital Nomad.
4. The Local.

Digital Valtellina

Digital Valtellina is a regional strategy for remote work. Our vision is to revitalize Valtellina's local economies and develop tourism through remote work. This is centered on the following principles:

1. Start up a "Remote Work Valley".
2. Build a new knowledge economy.
3. Develop sustainable tourism.

To start up a "Remote Work Valley", we propose that Valtellina build a strong brand and distinct regional identity as a hub for remote workers. This includes setting up the digital and physical infrastructure that will support and enable remote work (e.g., broadband internet, co-working spaces, cafes) and extending out Milan's business ecosystem to Valtellina (e.g., business partnerships, better transit connectivity) (Fig. 11).

Fig. 11 Start up a "Remote Work Valley". Source: image by the authors

To build a new knowledge economy, we propose setting up community hubs to enable knowledge exchange between remote workers and locals, creating training centers and programs for local skill-building (e.g., language, digital tools), and promoting local businesses and entrepreneurship (Fig. 12).

Last, to develop sustainable tourism, we propose to center the valley around sustainability. This includes promoting long-term engagement with the community, as well as building on the natural, historic, and cultural landscape of the valley (Fig. 13).

This regional strategy includes physical, digital, and policy interventions (Fig. 14) which are described below in more detail.

Physical

For physical interventions, we developed a place-based framework where we categorize the places across the valley based on their potential for remote work into city-town-village typologies (Fig. 15). We envision this as a slow, sustainable route

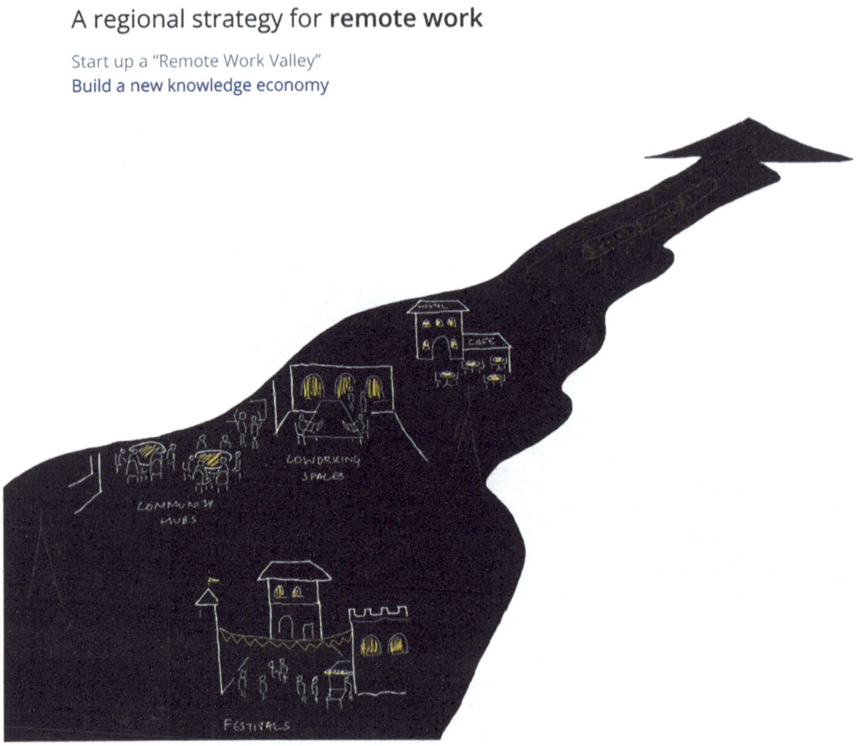

Fig. 12 Build a new knowledge economy. Source: image by the authors

that stitches together places and experiences across the valley in different ways: along the train and bike paths in the valley, as well as through the hiking paths and vineyards in the hillsides.

In this framework, we see Milan as the megacity (Fig. 16), both as a global entry point to Italy and the largest city close to the valley. The valley, with its passage to Switzerland, serves as a gradual exit from the big city into nature. This proximity to Milan could help build strategic partnerships with working hubs in Milan and create opportunities for short-term getaways from the city for both locals and visitors. The train line that connects to the historic Bernina Express at Sondrio provides an easy way of getting to different locations in the valley and our proposal places strong emphasis on the train connection.

Sondrio, the provincial capital, is the city in the valley. With a population of over 21,000 residents, it has an entrepreneurial local economy with a mix of old banking and hospitality industries and newer co-working hubs, culinary experiences, and seasonal winter games accommodations. We envision this city typology as the entry point to the valley, offering diverse living, working and local engagement options for remote workers while expanding local entrepreneurship and the knowledge economy (Fig. 17). Sondrio currently has traditional hotels for tourists and some

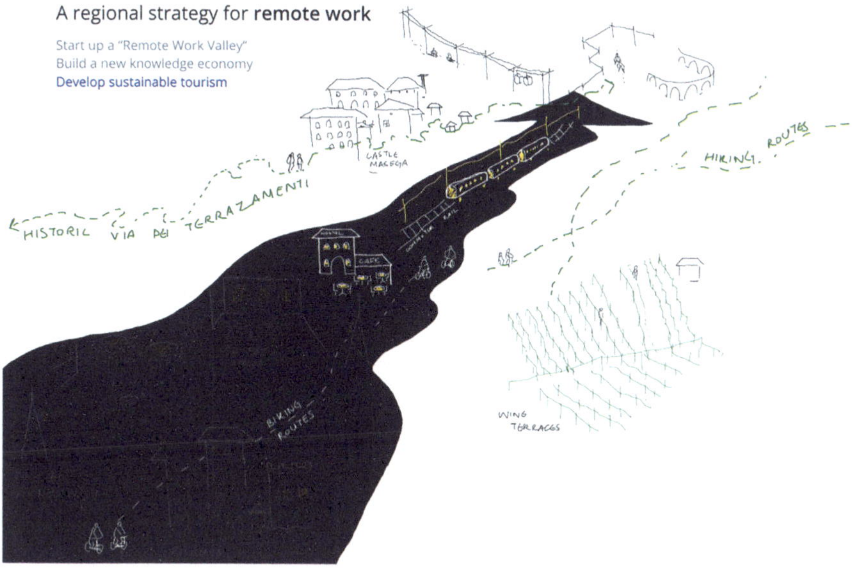

Fig. 13 Develop sustainable tourism. Source: image by the authors

newer hostels for sports enthusiasts. We propose expanding these existing typologies to include homestays, bed and breakfasts (BnB's) and boutique hotels for the visitors. We also propose the development of community hubs that would help remote workers engage with the locals. These community hubs could be combined with co-working spaces and cafes to work out of and restaurants for culinary experiences.

We identify towns as mid-size settlements in the region that have unique cultural offerings such as the Bernina Express in Tirano and the skiing areas in Livigno and Bormio. We envision the towns to act as hubs for sustainable tourism and local experiences for the visitors while leveraging these cultural offerings (Fig. 18). There are new eco-hotels in Tirano which offer sustainable living experiences and local homestays which offer opportunities for interactions with locals. We propose the development of community hubs with working spaces and food options. Tirano also has local restaurants offering high-quality local seasonal produce-based meals. The location of the towns along the valley offers easy access to the biking and hiking routes for the remote workers seeking nature-based activities. To leverage the seasonality and promote local culture, we propose highlighting seasonal festivals (such as Tirano's Eroico Rosso Sforzato Wine Festival), organizing remote work meet-ups and winter sport events in and around these towns.

Our final typology, the village, caters to remote workers seeking a rustic experience off the beaten path (Fig. 19). The villages are located on a higher elevation to from the valley floor and have a close relationship with the surrounding landscape. One of the villages along the Via dei Terrazamenti, Teglio, is located on a small hilltop and offers panoramic views of the valley. It houses the Accademia del

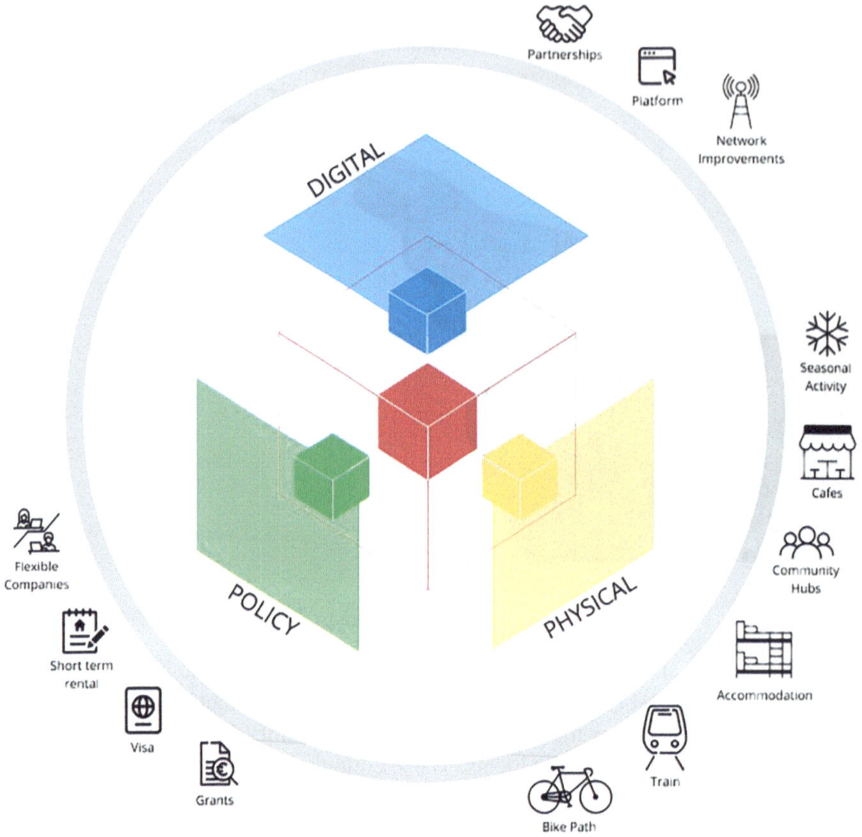

Fig. 14 The strategy for Digital Valtellina. Source: image by the authors

Pizzocchero, a local culinary school where one can learn how to make the local buckwheat pasta. The accommodation options in the villages include farm stays, family-run B&Bs, and homestays closer to the vineyards.

Last, we propose strengthening infrastructural development across the valley to facilitate a cohesive experience of the region. We envision the historic terrace route, the biking trails, and the train line that run through the valley as the main arteries for sustainable transportation and movement. We also focus on the development of high-speed internet infrastructure in the valley to support both remote workers and development of a local knowledge economy (Fig. 20).

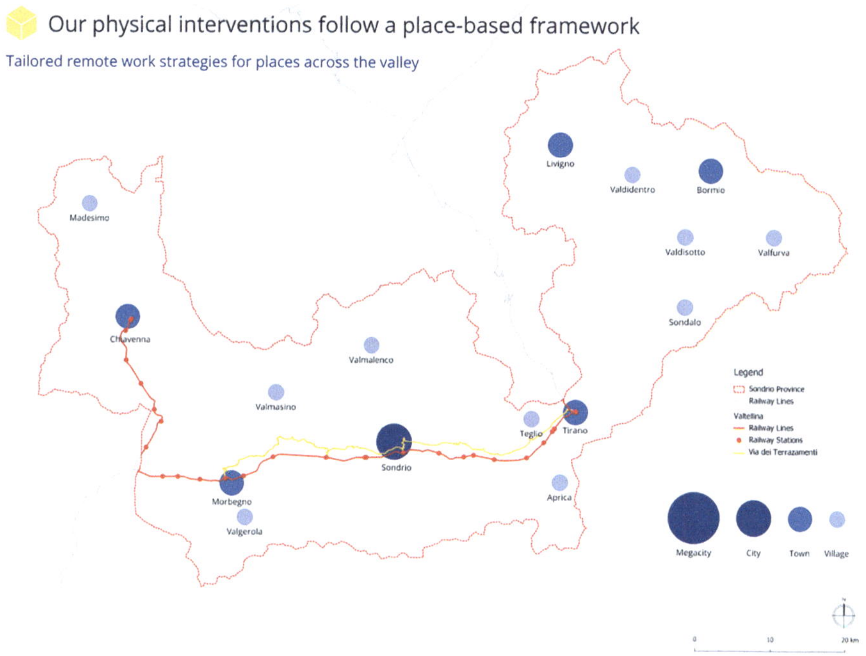

Fig. 15 A placed-based framework for the valley. Source: image by the authors

Digital

For digital interventions, we first propose to create a platform for remote workers to engage with Valtellina before they decide to visit. This includes an interactive 3D model to showcase city-town-village experiences, which allows visitors to get a feel for the place. It also has basic information about the place (e.g., housing options, internet speeds, cost of living), activities, and volunteering opportunities to help visitors find the specific type of experience that they are looking for.

Second, we propose a digital marketing strategy to promote Valtellina. For example, this can involve partnerships with major travel companies (e.g., Expedia, Airbnb) to showcase the region and direct traffic to the digital platform. It might also involve social media campaigns through various platforms (e.g., TikTok, Youtube, Instagram) or going through the Milan ecosystem of hotels and co-working spaces.

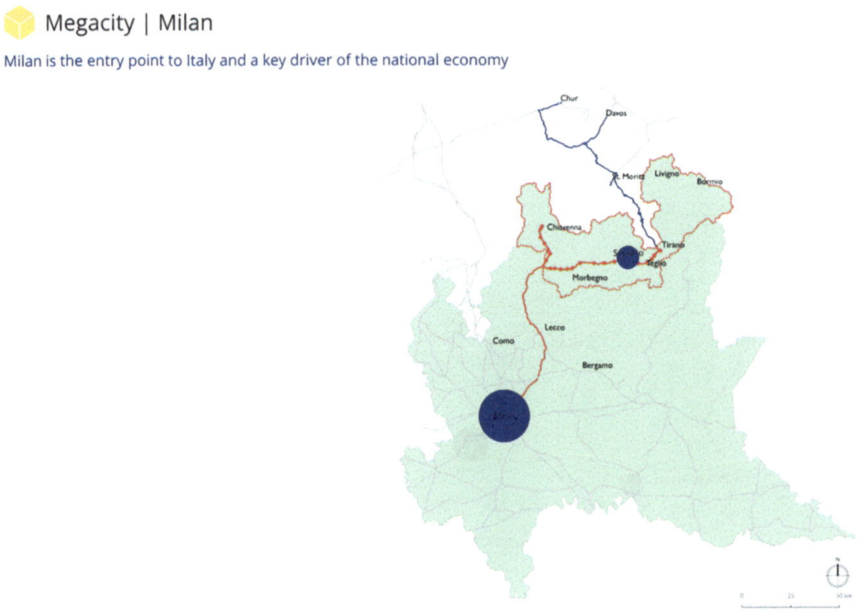

Fig. 16 Milan is the megacity. Source: image by the authors

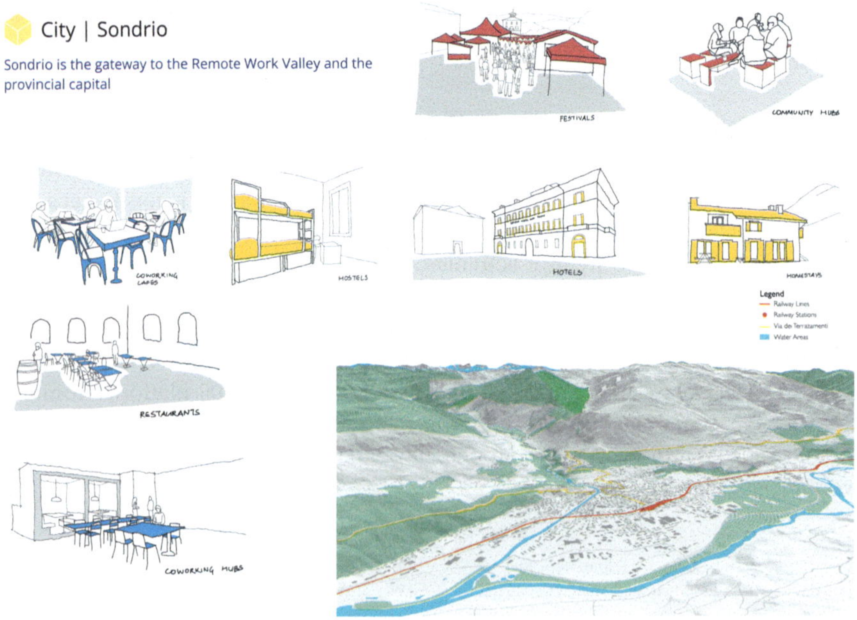

Fig. 17 Strategies for Sondrio, the city in the valley. Source: image by the authors

Fig. 18 Strategies for Tirano, a town in the valley that offers a local cultural experience and a connection to the Bernina Express. Source: image by the authors

Policy

Our strategy suggests policy interventions at the national, regional, and city scale (Fig. 21). These policy recommendations aim to support and further the overall strategy proposal.

On the national level, we propose a remote workers' visa to support the expansion of this proposal to attract global remote workers. There are more policy opportunities at the regional level. This includes creating tax benefits for remote workers who work from the valley, establishing partnerships with remote work-friendly companies to bring employees to Valtellina, and setting up grants for community hubs, co-working hubs and cafes to improve digital connectivity. At the local level, this includes creating additional grants to support local entrepreneurship endeavors and skill-building, as well as revisiting short-term rental regulations to allow for more accommodation options for remote workers.

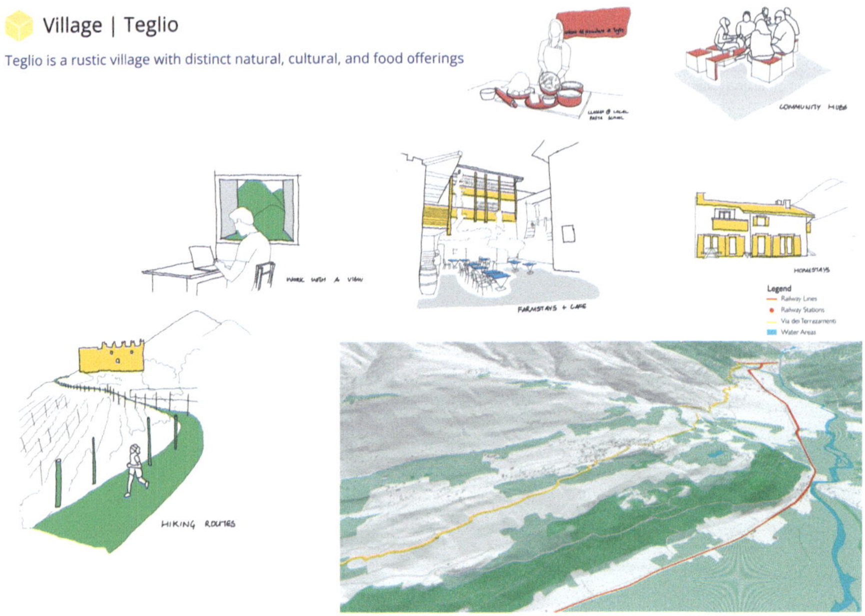

Fig. 19 Strategies for Teglio, a historic village on a hilltop in the valley. Source: image by the authors

Implementation

This strategy involves many stakeholders for implementation (Fig. 22). First, the regional government must get buy-in from all stakeholders. It needs to drive the strategy forward and implement on it. As part of this, the regional government must partner with local stakeholders (e.g., cities, local enterprises, communities) to scale up capacity for remote work. This includes using regional grants to fund physical and digital infrastructure, as well as maintaining close dialogue with local enterprises and cities to make policy changes.

Beyond this, the regional government must work with the national government and global businesses to attract remote workers from around the world. This includes making policy interventions (e.g., remote workers' visa, community grants) and digital investments (e.g., network improvements, digital outreach, business partnerships).

In terms of timing, this implementation has two phases. The first phase focuses on building out the digital and physical infrastructure and policy needs to meet minimum requirements for tourism and remote work, generating momentum, and crafting a distinct identity for the valley. This targets the tourist and hybrid worker personas.

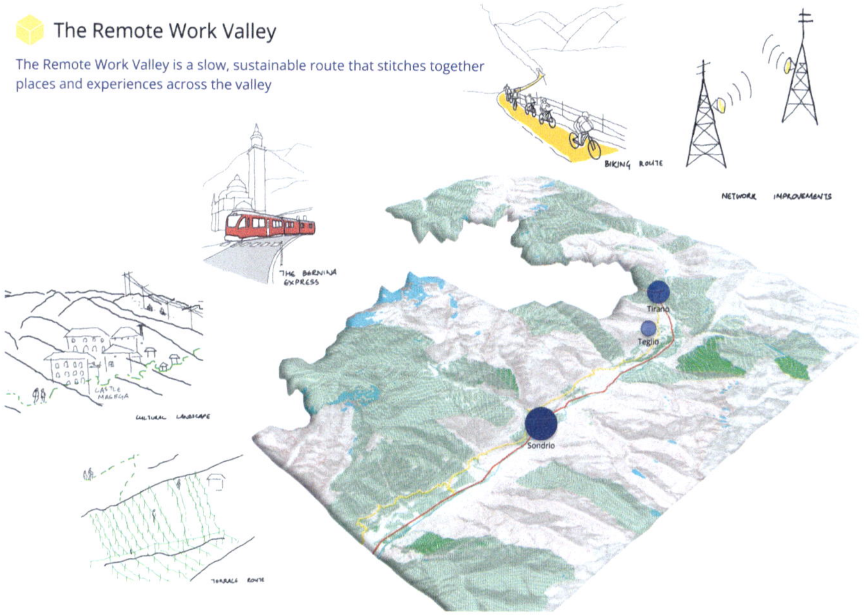

Fig. 20 Strategies for the Remote Work Valley. Source: image by the authors

Fig. 21 Policy recommendations to attract remote workers. Source: image by the authors

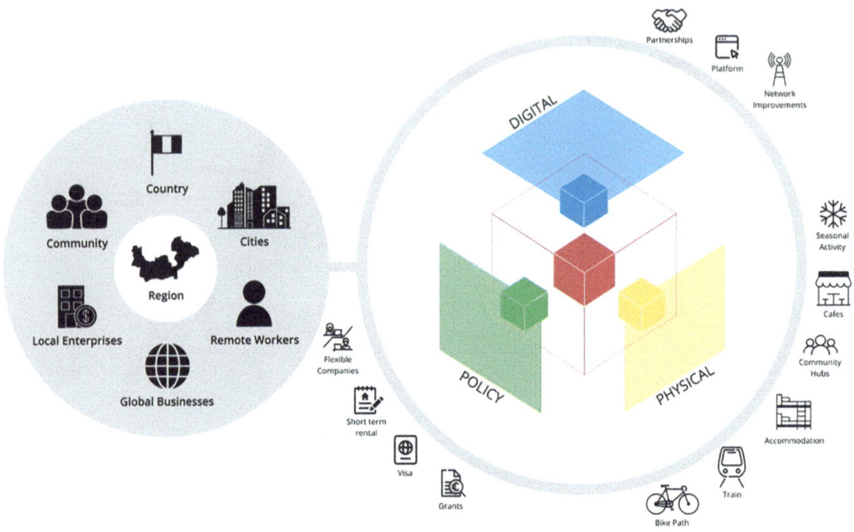

Fig. 22 Implementation strategy. Source: image by the authors

The second phase involves setting up community hubs, training programs, and building up other local initiatives to sustain the new knowledge economy in the long-term. This sets up Valtellina to fulfill the needs of digital nomads and locals.

Personas

To bridge this strategy and its implementation to the personas that we developed through our engagement with remote workers, we built some narratives around how we imagine various personas, the Hybrid Worker and the Digital Nomad, to experience Digital Valtellina.

The Hybrid Worker

The Hybrid Worker works in Milan (Fig. 23). She hears about Valtellina's sustainable tourism through her co-working space and signs up for a pasta-making class. She takes the train to Valtellina for a long weekend to take the class and works from a co-working café. She falls in love with the slower pace of life and comes back to discover more: hiking, biking, and wine tasting. She sees the potential in Valtellina and decides to mentor young entrepreneurs in the valley.

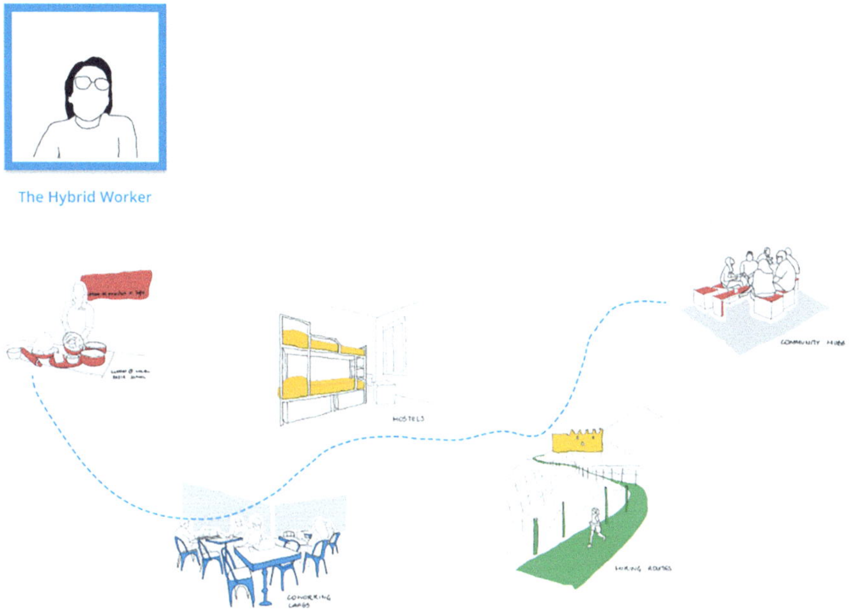

Fig. 23 The Hybrid Worker. Source: image by the authors

The Digital Nomad

The Digital Nomad is thinking about her next destination (Fig. 24). She's been learning a bit of Italian online and wants to get immersed in the language. She has visited Italy before and traveled through Valtellina for the Winter Olympics. She has great memories from her trip and wants to revisit for a longer time. She finds homestays on the digital platform and sees that internet speeds have improved. She is surprised to find that there are a ton of events and new local businesses. Sondrio checks all the boxes, and she decides that Valtellina is her next destination!

Conclusion

Cities around the world are adapting to the challenges of remote work. Digital Valtellina explores how Italian's inner areas can prepare for and manage this digital transformation. Here, remote work can be leveraged as an opportunity to revitalize Valtellina's small towns and promote sustainable tourism. The project aims to address the challenges posed by remote work and envision a transformation for cities in response to this trend. During the project, we examined how cities should adapt and respond to remote work, questioning whether it can be used as an advantage rather than a struggle with empty offices. We explored the potential for cities to

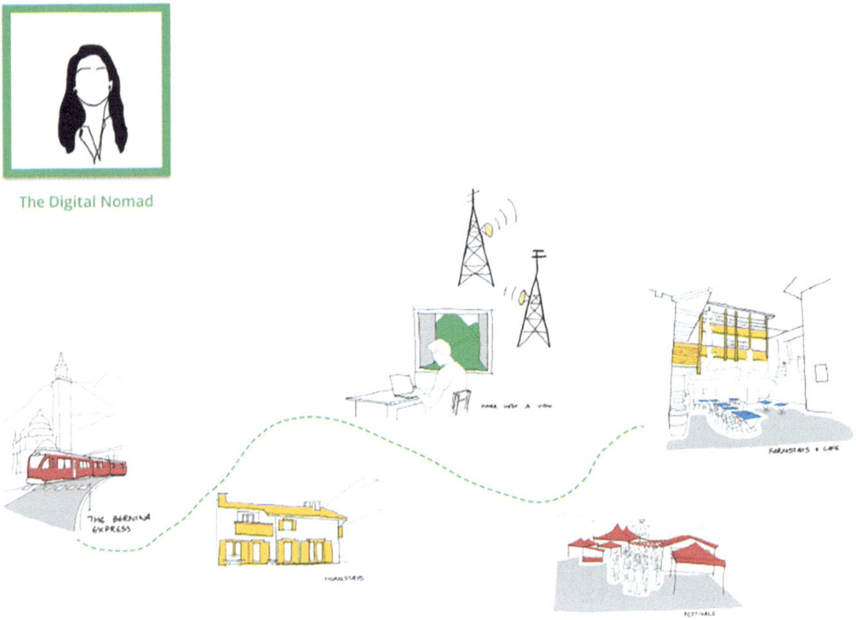

Fig. 24 The Digital Nomad. Source: image by the authors

reverse the narrative and explored how remote work can contribute to their development. Inspired by the shrinking towns we observed in Italy and the rise of remote work during the pandemic, we dedicated weeks to testing, developing, and refining our idea.

The outcome of our project is a regional strategy designed to attract remote workers to Valtellina. This strategy encompasses physical, digital, and policy-based approaches, and we provide examples of these strategies, identify the stakeholders involved, and outline the implementation process. The strategy is the culmination of our research and engagement throughout the semester.

Upon presenting our strategy to the Sondrio municipal authorities, we discovered that they were aligned with our proposed interventions. The main strength of our proposal lies in the combined vision for Valtellina as a slow-sustainable remote work valley. This vision helps to connect the various interventions and fill gaps in the current plan. It presents a bold idea that stretches the imagination about what Valtellina could look like and offers new possibilities for the region. It is highly relevant and builds upon the growing discourse surrounding the future of remote work in cities. We have developed a comprehensive range of physical, digital, and policy strategies that can be employed by the region to attract remote workers. By addressing various aspects of the remote work ecosystem, we ensure a holistic and well-rounded approach to support the growth of remote work in Valtellina. We advocate for a phased approach, starting with tourism and hybrid work as initial steps before fully embracing remote work. This incremental strategy allows for a smoother transition and ensures the region can effectively leverage its existing strengths and resources. The proposal is well-aligned with existing initiatives in cities like Sondrio. We have considered the ongoing efforts to create a "Smart City" in Sondrio and the broader regional plans for a sports and sustainable mobility corridor through Valtellina. By incorporating these existing frameworks, we maximize synergies and create a cohesive vision that can be effectively implemented.

Reference

Micallef, J. V. (2018). The Valtellina: Discovering Nebbiolo from the Alps. *Forbes*. https://www.forbes.com/sites/joemicallef/2018/07/14/the-valtellina-discovering-nebbiolo-from-the-alps/?sh=2e92997f3864. Last accessed 8 Nov 2024

The Smart Terrace Trail as Strategic Regional Project

Giovanna Fossa, Felix Günther, and Deborah Paolini

Vision

Recently, especially after the covid, historical walking routes ("cammini") are emerging as rebooting strategy for shrinking rural areas in Europe, thanks to their potential in terms of contemporary trends for experience tourism and holistic well-being. In parallel new telecommunication technologies are introducing a digital accessibility which can overcome physical marginality and orographic barriers providing a high-level connectivity even to remote villages. This scenario allows us to develop an updated interpretation of the walking route as linear platform combining working and vacation for innovative lifestyles. It is a new concept framed in a regeneration vision for historical shrinking landscapes leveraging more on strategic bottom-up projects than on top-down policies. In this perspective we design a smart evolution of the Terrace Route in the heart of the Valtellina, proposing it as pilot project for the regeneration of mountain contexts that are below the elevation required to ensure enough snow for skiable domains, mild valleys which are the most fragile typology of shrinking territories in Europe.

G. Fossa (✉) · D. Paolini
Department of Architecture, Built Environment and Construction Engineering, Politecnico di Milano, Milan, Italy
e-mail: giovanna.fossa@polimi.it; deborah.paolini@polimi.it

F. Günther
Department of Environment, Constructions and Design, University of Applied Sciences and Arts of Southern Switzerland, Mendrisio, Switzerland
e-mail: felix.guenther@supsi.ch

G. Fossa et al. (eds.), *Small-Town Renaissance*, Research for Development,
https://doi.org/10.1007/978-3-031-89786-3_8

The Regeneration Potential of Walking Routes for Shrinking Rural Landscapes

Walking routes are itineraries along historic paths of great landscape value, crossing orographically diverse territories and containing ancient settlement nuclei: historic, productive, integrated and authentic landscapes that have been cut off from the main axes of development by the major infrastructures since the beginning of the twentieth century, and whose rural and artisanal economy, although of high quality, is being undermined by contemporary globalization processes. These areas are also threatened by the counterproductive effects of energy and food policies at continental level, which disregard local specificities and the heritage value of these landscapes (e.g. panel fields instead of cultivations, etc.), and even by environmental policies that sometimes go too far in re-naturalizing historical and heavily anthropized areas (e.g. the controversial issue of bears in Trentino or the recent proliferation of wolves in the Alps and the Apennines, which are damaging pastures). Italian policies refer to these areas as 'inland', an adjective that highlights the distance from the coastline, and intuitively describes the depopulated inland areas of islands such as Sicily and Sardinia and the mountainous contexts of the Apennine ridge, but poorly describes the Alpine depopulated areas, which are often border areas.

Not only Italy but the entire European continent is furrowed by these 'ways' of medieval origin, pilgrims' routes between historic cathedrals, through man-made rural landscapes formed during millennia. The most famous are the Way of St. James, which crosses the Pyrenees massif and the mountains of Asturias to reach the Atlantic in Galicia, the Via Francigena, which runs along the Italian Apennine ridge, and the Russian pilgrim's route, which reaches the Mount Athos peninsula in Greece. In addition to these main routes of medieval religious origin, the network of historical walking routes includes ancient commercial routes or rural connections between valleys: transalpine and trans-Apennine routes in Italy, and a whole series of minor routes between villages perched on hillsides and agricultural landscapes of scenic value (such as the paths of transhumance spread between pastures and plains in central and northern Italy, the famous itineraries of the "5 Terre" in Liguria, and those more recently secured and arranged for tourism purposes, such as the Prosecco trails in the Veneto pre-Alps and the "Via dei Terrazzamenti" in Valtellina, subject of this project proposal). Historical paths of military or defensive origin are also to be mentioned, such as the Cadorna path in the Alps or the "Camin de Cavalls", surrounding the island of Menorca, that has provided the interpretive key to relaunching its touristic image in an identity-based manner within the Balearic Islands system.

European trails are itineraries of one or more days, or even weeks, that allow an authentic encounter with a landscape that is very rich in history, a territory that is still alive and productive, not musealized, an experience that has cultural, sporting, existential and spiritual values. They are fundamentally different from the American Nature Trails (such as the Continental Divide and the Appalachian Trail, etc.), which take their inspiration from wilderness.

The rediscovery of the 'paths' has occurred as a result of their potential for experiential tourism, which represents one of the most significant trends in contemporary tourism. From this perspective, the ability of experiential tourism to trigger regeneration in entire territorial contexts is particularly noteworthy.

Furthermore, this process does not adhere to external models; rather, it is based on the reinforcement of the historical identity of these territories and of the local communities that continue to inhabit them; indeed, despite the richness in cultural heritage, the risk of a loss of identity is emerging as a result of the decline of local communities due to depopulation phenomena. This risk of cultural heritage is to be considered in addition to the other risks typical of marginal territories (Foppoli, Menghini and Zanolari, 2014; Moioli, 2023).

Experiential tourism can be a means of revitalizing rural contexts that are marginal but rich in artistic, architectural, historical, cultural assets. These contexts often combine an excellent food & wine based on local organic agriculture and a valuable landscape with story-telling products. Examples of such landscapes include vineyards, following the example of Tuscany, or crops that give rise to blossoming (e.g. Castelluccio di Norcia in the Apennines). Indeed, experiential tourism can be sustainably integrated into local communities and economies, enhancing their specific characteristics. Moreover, contemporary art and architecture could drive regeneration processes of shrinking landscapes, often involving traditional artisan skills in the art production. Among many success stories of hill-mountain villages re-born through art, especially if supported by being located on historical paths, we could cite the pioneer case of "Arte Sella", conceived in 1986 in a hidden valley of the Trentino, the one of Illegio in the Carnia mountains, where the local community has been hosting art events since 2004, and the more recent example of OCA Oasy Contemporary Art opened in 2023 in Tuscany along the via Francigena, combining contemporary creativity with recreational therapies for handicapped people.

Low-altitude mountain territories are particularly at risk, squeezed between the more and more attractive lakes at the foot of the Alps (above all, but not only, the Como Lake) and the famous ski areas at high altitudes (Varotto, 2020). Their valley floors are being ruined by industrial and commercial conurbations allowed by top-down planning and policies, even in presence of built heritage and landscape quality cultivations on the surrounding slopes. In the face of growing urbanism, the ongoing change of the rural development model[1] shows unprecedent "glocal" synergies and tensions: crop industrialization & real estate pressure versus organic farming & agro-tourism, at the front-line to counter the abandonment of inner rural areas.

The mountain tourism development model seems to be under discussion. In the Alps and the Apennines in particular, but in Europe in general, the extension of downhill skiing to the masses requires for its further development large, interconnected ski areas at higher and higher altitudes (ideal altitude above 1500–2000 m), in order to ensure natural snow coverage despite climate change. The development

[1] Cf. the exhibition "Countryside the Future" by R. Kooolhaas at the Guggenheim NY in 2020.

of winter sports tourism in the Alps and the Italian Apennines, which commenced in the 1970s, has resulted in the construction of individual ski lifts at altitudes even lower than 1000 m. It is evident that this model of development has reached maturity and is now in a state of decline.[2] The consequence of this is the widespread presence of disused ski infrastructure, or their continued operation with the use of artificial snow, which is created thanks to public funds. Alternatively, these ski lifts may be kept open for the summer season, with focus on activities such as mountain biking or hiking.

The approach typical of the 1980s, exemplified by the 'great events' such as the World Cup and the Olympics, has also demonstrated its limitations. Indeed mountain communities have often been subjected to decisions made by national stakeholders, which pushed the construction of large facilities and infrastructure keeping the occasion of these events.[3] Such projects have frequently resulted in the jeopardy of centuries-old natural and landscape resources. In many cases this approach caused the destruction of the very source of the tourist appeal, as well as the erosion of the rootedness of the tourist offer in local traditions and identities. This has led to a globalization of the tourism model.

Swiss regions have recently turned down Olympic and World Cup bids through popular votes, reflecting public concern over environmental impacts and costs. Notably, the Canton of Grisons rejected Winter Olympic bids twice in 2013 and 2022 and Valais voted against a bid for the 2026 Winter Olympics, confirming the trend in Switzerland where local communities prioritize environmental and feasibility considerations over hosting large international events.

The promotion of heritage by UNESCO has brought attention to some of these depopulated and unfashionable territories through rebranding operations that were undoubtedly successful but not without their shortcomings. These include a lack of stimulus to local entrepreneurship and the risk of inflation resulting from the multiplication of such declarations, the overexposure of limited areas of excellence without adequate contextualization and dialogue with local communities, as evidenced by the recent reactions from environmentalist associations and Italian Alpine Club[4] against the Dolomiti UNESCO declaration,[5] which only encompassed the monumental peaks zone.

While climate change is bringing the habitat of the low-altitude mountain closer to that of the countryside (with even the presence of mosquitoes now reaching 1500 meters, albeit statistically less significant), on the other hand, relatively higher altitudes, even if they do not guarantee snow, offer less humidity, more ventilation and,

---

[2] In the current climate-change framework, the 1000 m level could be rediscovered as the best altitude for sport training and rehabilitation. Cf. the example of Campodolcino in the Spluga valley.

[3] Cf. the national commissioning of the Cortina local government as far as related to the Olympics, and in particular the controversial construction of the bobsleigh run.

[4] See the interview to the CAI President in the *Corriere delle Alpi* newspaper, August 26th, 2024 and the official statement by the non-profit organization *Mountain Wilderness*, September 2nd, 2024.

[5] The Dolomites were declared UNESCO World Heritage in 2009.

in terms of perception, milder winters and summers. Real estate investors are beginning to pay some attention to these quality of life indicators.

The pandemic has contributed to the promotion of "walking", because of the associated health benefits (physical and psychological), and to the emergence of new outdoor lifestyles that value the integration of settlement and open space, typical of historic landscapes. Walking has been scientifically proven to be good for health (heart, bones, obesity) and is now part of mainstream therapeutic practice, preventing the health risks caused by sedentary lifestyles, which are increasing exponentially in today's urban environment. Recent studies claim that incorporating walking into work rhythms improves performance and refreshes and clears the mind; walking also supports local small-scale economic activity through spending along the way, fits in with zero-kilometer economic trends, facilitates social inclusion and a sense of belonging to a community.

Walking increasingly integrates other functions, becoming a multitask activity and creating neologisms: *FastWalking*, *FitWalking (*a body therapy, beyond sport-inspired "vita-parcours"); *SmartWalking* (smart-working integrated with walking, supported by outdoor workstations, outdoor coworking spaces, diffused plug-in and continuous WIFI[6]); *Walking Counseling* (strategic counseling integrated with walking[7]); *Deep Walking* (training and meditation, psychotherapeutic activity, a re-edition of pilgrimage walks, with yoga or mindfulness approaches)[8]; *Walking Teaching and Learning; Walking Workshops.*

On this basis, walking intercepts the change in mentality and imaginary that is taking place in parallel with the revolution of living/working styles. A new paradigm of walkability is emerging, with multitask walking spaces: a system of public spaces and services designed primarily for walking as a therapeutic, working and training activity. In this new concept, essential services are found "integrated" in the walking activity itself (beyond the requirement to reach them within 15'). Urban and territorial walking spaces are envisioned seamlessly: the "human scale" territory emerges from urban walkable domains and embraces the whole region, in a resilient and attractive way. The scale of the future for pedestrian mobility is large, linking historic centers to regional walking networks (trekking, hiking). Climate change is increasing the proportion of year-round pedestrians.

At the wide area scale, the focus has been primarily on bicycle greenways, while prioritizing pedestrian routes changes the design criteria and the type and frequency of services to be correlated (widespread digital coverage, outdoor eating and coworking, agritourism, local markets, viewpoints, land art, visual and performing arts, on-demand drones, etc.).

Walking routes are thus seen as strategic structures that integrate synergistic roles: destination for experiential tourism and contemporary grand tours as an

---

[6] Cf. the blog *Smart Walking* by Davide Fiz, 2021, and also OCD *Outdoor Coworking Day* Events.

[7] Cf. Serenella Sala & Associates, Executive Counseling Services, Milano: *Walk the Change* Program.

[8] Cf. for example Alessandra Maffi, Mindful Coaching Services, Bergamo, who proposes therapeutic walking along the Adda River.

alternative to global mass tourism; garrison against the depopulation of inner areas; valorization of historic productive landscapes; regeneration drivers of the conurbated valley floors.

The public space of the future can be conceptualized as a landscape continuum that supports a polycentric regional development through the whole range of sustainable mobility ways (on foot, bicycle, rail, boat, and cable car). This approach allows a development vision which integrates various scales, enhancing landscape identities.[9]

At the international level, design experiments are inspired by new imaginaries of pedestrianism, which outline resilient, multifunctional and multi-scalar walkways.[10] These walkways often regenerate former infrastructural spaces and integrate ecological networks in terms of biodiversity and climate change adaptation, improving microclimate and land drainage.

In addition to these sociological phenomena and experimental projects, the current digital transition makes it possible to overcome physical marginality with new technologies that connect with professional download and data exchange speeds, even in places that cannot be reached by fiber optics, such as dispersed rural settlements or slopes.[11] The technology overcomes orographic difficulties and provides digital accessibility to contexts that are poorly served by the mobility network. This opens up unprecedented possibilities for smart and remote working, as well as for visiting cultural assets that are dispersed in inland areas (Dania, Gazzerro, 2023). New strategies for resilient polycentrism become possible, developing what was temporary experienced during the covid (the rediscovery of underutilized family homes or second homes in rural and mountain locations, thanks to remote working).

Furthermore, experiential tourism meets with digital nomadism; this represents a vision of year-round tourism that is integrated with the territory. New e-learning facilities allow young people to continue their studies inside the native context, even at the university level.[12] The aim is to create a situation whereby territories are once again places where people live, work and produce goods and services, rather than being abandoned or offered for seasonal consumption.

It is crucial to underscore that a territorial model for promoting tourism must not be devised solely based on its appeal to visitors, their itineraries, and their activities. Furthermore, the sustainable development of settlements and mobility, supply and disposal for the resident population must be considered (Schurch, 2006). Unfortunately, this is often in conflict with the valorization of landscape quality for tourism. One aspect of the regional landscape and culture that should not be

[9] Cf. Fossa G. and Günther F. contributions at the Conference "Baukultur Ticino 2050 Scenari", SUPSI Mendrisio, March 4th, 2024.

[10] Examples of innovative walkways include the NY/NJ waterfront from *Little Island* to the *Meadowland Resilient Park*, and the dry/wet trails of Norway.

[11] cf. FWA, Fixed Wireless Accessibility; high and low flying satellites etc.

[12] Politecnico di Milano opens in the A.Y. 2025/26 an innovative e-learning center in Sondrio (Valtellina).

underestimated is its impact on the attractiveness of the area for local communities, including residents, workers, and for their leisure time.

A Pilot Project: Making the Terrace Route "Smart"

Project Context: The Valtellina Region

The proposed pilot project[13] concerns the middle Valtellina, a typical low-altitude mountain context, both for risks (flooding and landslides, depopulation, abandonment of pastures and wood, loss of craftsmanship activities and skills) and potentials (cultural and built heritage, historical rural landscape). The current situation is dominated by the valley floor conurbation and the decline of upper villages and lateral valleys. This territory has to be interpreted in the frame of the whole Valtellina, at the heart of the Alps, at the border between Italy and Switzerland.

The Valtellina is a *landscape of man* since prehistoric times; its thermal waters were well known by the Romans; in the Middle Ages the valley was the continental crossing point of wine and salt paths; however, at the beginning of the XX century, the development of the modern European infrastructure corridors (with trans-Alpine tunnels for road and railway) selected different axes (Gothard and Brenner), bypassing Valtellina.

The valley is difficult to access and also digitally not yet well connected (ongoing upgrading programs), despite its proximity to the Milan metropolitan area (the main center, Sondrio, is 130 km away from Milan). Valtellina is a long dead-end valley served by a main road in the bottom of the valley, parallel to a one-track railway which interchanges in Tirano with the Swiss train network. Most of the mountain passes connecting with the surrounding valleys are closed in winter because of snow and avalanches.

It is grooved by the Adda river, tributary of the Lake Como, and belongs to the great Po River watershed. In 1987 the big landslide of S. Antonio Morignone (in the upper valley) and the consequent flood of the Adda river had disastrous effects on the whole valley, making Valtellina an icon of mountain settlements at risk. This tragedy made Valtellina also an experimental laboratory for public policies, with alternating effects. In this framework has to be seen also the decision to offer to upper Valtellina some events of the Olympics 2026. Its shrinking inspired the current regional support for living in little mountain villages.

---

[13]The proposed pilot project has been developed in the framework of the *Regional Planning Design Lab* led by Proff. Fossa and Gunther, Politecnico di Milano (A.Y. 2022–2023) in synergy with the *Valtellina Workshop* "Towards the Smart Villages of Italy" organized by Milan Politecnico/MIT Progetto Rocca/MISTI Italy (September 2022). The participatory process which accompanied the project is summarized in the chapter Boniotti, C., Fossa, G. "Stakeholder Perspectives for Regenerating Inner Territories".

The middle Valtellina looks west towards the tip of the Lake Como (altitude 210 m above the sea level), proceeding eastwards to the upper valley which represents a traditional Milan touristic destination for winter and summer (with villages located at 1400–2000 m). Main centers are (from West to East) Morbegno, Sondrio and Tirano.

Morbegno (12,000 inhabitants, 260 m a.s.l.), where the minor lateral valleys of Valmasino and Valgerola merge, boasts an interesting historic center and the famous cheese "bitto".

Sondrio, the Valtellina main town (25,000 inhabitants, 300 m a.s.l.), capital of the province, which service economy is suffering local banks acquisition by multinational groups and local hospital closures for concentration in major regional health poles; located on the Valmalenco dejection cone, presents little villages on the overhanging slopes, some of them abandoned.

Tirano (9000 inhabitants, 450 m asl) is a cross-border node. It is the terminal of the Italian railway interchanging with the Bernina Rhaetian railway (historical infrastructure which is recognized as UNESCO heritage). It is located at the crossing point between Valtellina, the Swiss Valposchiavo (North) and the Val Camonica (South, towards Brescia). Its landscape is unique for the integration (morphological and functional) of the rich historical nucleus with the river and the vineyards on the surrounding slopes. Ancient paths connect the Tirano historic center to the terraced vineyards.

The middle valley is a historical rural-productive landscape, which still now preserves at a certain level the authenticity of a year-around living/working settlement. This feature is in line with the contemporary experience tourism demands.

This selected stretch of Valtellina highlights the shape of the structure of the valley settlement system, which combines the recent linear conurbation of the valley floor with a sequence of historical villages halfway up on the sunny slope (the one of the Rhaetian Alps). The slope is terraced by dry stone walls and cultivated with vineyards. The rural economy is at the origin of these settlement cores, which were born equidistant in order to preserve the vital agriculture space surrounding each of them. These villages are aligned along a historical road now called "The Scenic Road" ("Strada Panoramica", open to vehicles) which is intertwined with ancient paths across the vineyards, interconnected in a pedestrian and bike itinerary, called indeed "Route of Terraces" ("Via dei Terrazzamenti"). Actually, the historical backbone of the Valtellina settlement system is not the road in the valley floor but this road half-way the sunny mountain slope, with villages and vista points aligned across the vineyards (Figs. 1, 2, and 3).

Dry stone terraces sustaining vineyards are Unesco heritage since 2021 for the cultural value related to the traditional art of making them (Foppoli, 2019). Terracing here is coherent with the geological fault ("Insubric Line"), which shaped the Rhaetian side of the valley as a natural sunny wide "balcony", extending itself through terraced slopes, both upstream and downstream.

The typical historical transect of the middle valley shows high pastures for summer ("alpi"), medium altitude pastures for May ("maggenghi") and stables for winter in the valley floor. Halfway up the slopes, villages are surrounded by vineyards.

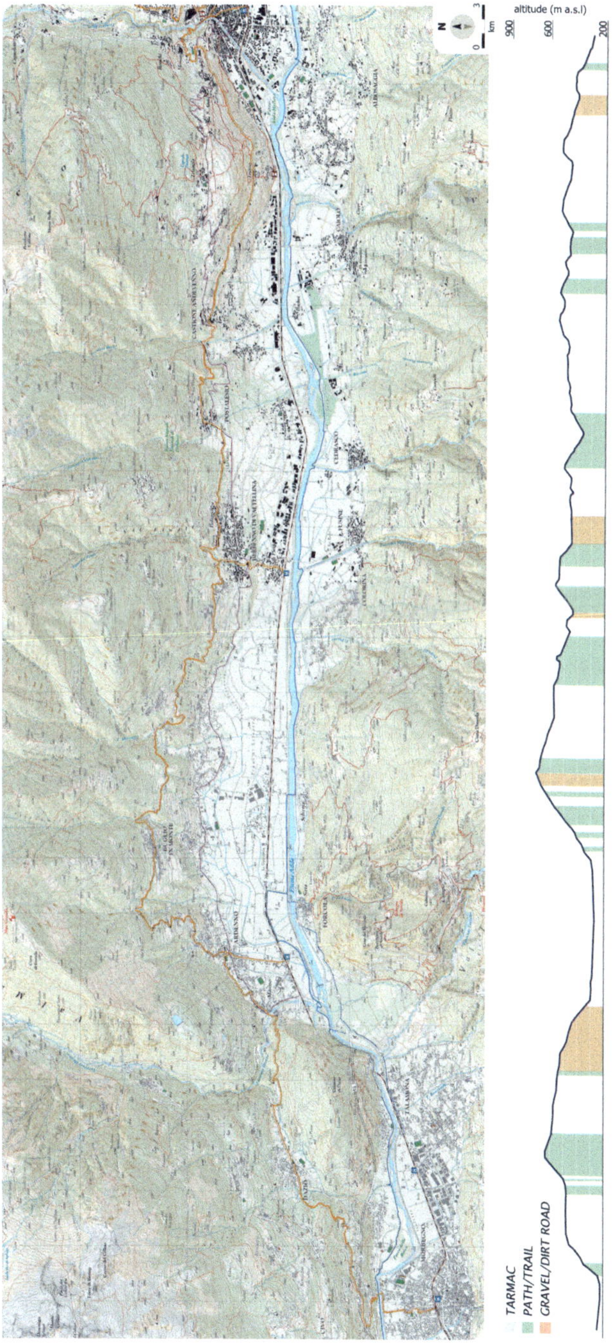

Fig. 1 Via dei Terrazzamenti ("Route of Terraces") from Morbegno to Sondrio. Source: graphic elaborations by Deborah Paolini on a cartographic image by Edizioni SeTeMap, SeTe srl—Poggiridenti SO

Fig. 2 Via dei Terrazzamenti ("Route of Terraces") from Sondrio to Tirano. Source: graphic elaborations by Deborah Paolini on a cartographic image by Edizioni SeTeMap, SeTe srl—Poggiridenti SO

Fig. 3 Walking along the Via dei Terrazzamenti during the workshop "Towards the Smart Villages of Italy", Milan Politecnico/MIT/MISTI, September 2022. Source: photo by Deborah Paolini, 2022

The upstream slopes still feature their pasture meadows and huts: some of these places are still in use (Pian dei Cavalli), others were reused in the '70s for vacation second homes (S. Bernardo, Boirolo) or as sport-excursion destinations (Alpe di Teglio, Prato Valentino). Very well exposed locations were selected in the early '900s for sanatorium buildings, now abandoned (Prasomaso).

The village of Teglio (4500 inhabitants, 900 m asl) gives the name to the whole valley (*Vallis Tellina*). It is located above a mountain pass shaped like a saddle. Teglio widely overlooks the valley landscape from its ancient watchtower. It is a focal point of the walking/bike path "Via dei Terrazzamenti".

The Rhaetian slope is furrowed by a series of valleys, often subject to hydrogeological risks. Some of them host dams for hydroelectric energy (Valmalenco above Sondrio, Val Fontana above Ponte, Valposchiavo above Tirano). The most important is Valposchiavo, which is grafted onto Tirano, connecting Engadina (Switzerland) with Valtellina.

The two valleys Valtellina and Engadina develop in parallel, with about 1500 m of difference in altitude (Fig. 4); Valtellina belongs to the Adda-Po watershed, Engadina to the one of the Inn-Danube, shaping a continental divide. S. Moritz is one of the first places to establish themselves as tourist, at the end of the XIX century, and to become world class destinations, thanks to the excellent accessibility from the Inn Valley and the barycentric location between the two metropolitan regions of Milan and Zurich, in addition to the perfect altitude for winter sports (above 1.800 m asl). Valtellina vs Engadina offers precious complementarities such as built and cultural heritage and food and wine attractions. Between these valleys there have been constant exchanges through Valposchiavo: the most important is represented by cross-border workers and also hydroelectric energy, but we could

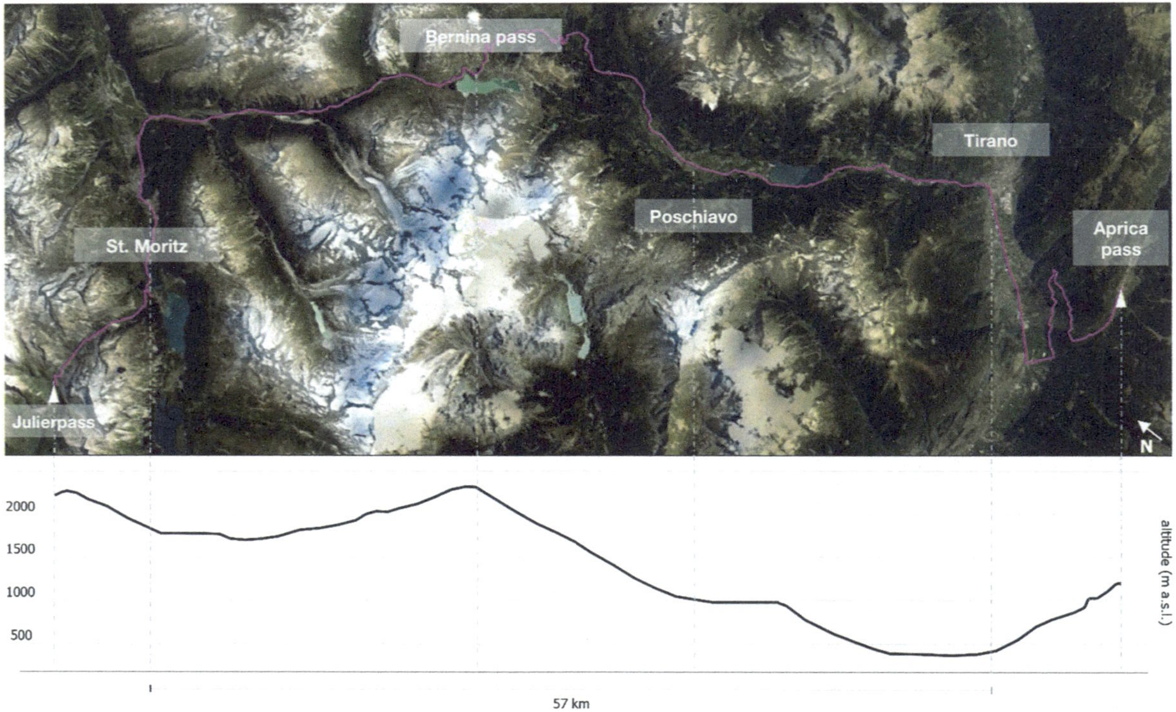

Fig. 4 Elevation profile along the historical mountain road connecting Tirano and S. Moritz and the side passes overcoming watersheds. It is evident the 1500 m difference in altitude between Engadina and Valtellina. Source: image elaborated by Deborah Paolini; basemap: Google Earth, Image Landsat/Copernicus 2021

cite also vernacular architecture cultural exchanges, reinterpreted in a modern language by the famous Milanese architect Luigi Caccia Dominioni.

Historical monumental houses feature the center of Tirano (Palazzo Salis, Venosta, Foppoli) as well as the cores of some villages along the Via dei Terrazzamenti (Palazzo Besta at Teglio and Bianzone). Villages are landmarked by these "Palazzi" or by historical churches and castles. The dismission phenomenon here is related to this typology of great families' mansions and castles, and also to early XX century mountain sanatoriums (Sondalo and Prasomaso); only marginally dismissions concern factories in the valley floor; the abandonment of rural houses and stables is diffused over the slopes, and we find many disused rural nuclei along the Terrace Route.

The opposite side of the valley, the Orobic slope, in the shade, is covered by woods: a nature reservoir in complementarity with the historical anthropization of the Rhaetian side.

A Problem-Solving Approach

For the project proposal to achieve its full effect, it must be embedded in its context, with a problem-solving approach. This applies to both its spatial functionality and its landscape quality. To this end, the key challenges facing the valley must be addressed. These are mainly identified as demographic issues, economic and tourism development, infrastructure and mobility, environment and sustainability, cultural and heritage preservation and valorization.

The demographic decline in Valtellina is one of the most pressing. The region has experienced significant population loss over the past decade, with young people migrating to urban centers like Milan and Zurich, resulting in a 2–4% population decline in the past 10 years.[14] This trend has left Valtellina with an aging population, as those who remain are predominantly older, leading to a shrinking workforce. The exodus of younger generations has left many abandoned villages and agricultural lands unattended. The so-called "brain drain", with only 8.9% of the population holding university degrees (compared to over 20% at national level)[15] is leaving the region with insufficient skilled workforce to support vital sectors such as agriculture, tourism, and technology and most notably to address the intricate developmental challenges that lie ahead.

A major issue is Valtellina's inadequate infrastructure, which hampers both local mobility and tourist accessibility. The main road connection SS38 suffers from traffic congestion during peak tourism seasons. These shortcomings in the transport network are said to limit the flow of both residents and tourists. Furthermore, Valtellina's high dependence on cars exacerbates these issues, with 85% of internal

[14] ISTAT, 2023.

[15] Data: Chamber of Commerce of the Sondrio Province, annual report, 2023.

travel conducted via private vehicles.[16] The lack of efficient transport options is due to the region's outdated public transport network. The Colico-Tirano railway line is said to be underutilized due to a combination of uncompetitive travel times, an infrequent train service, and an unstable timetable. Moreover, important tourism routes like the "Via dei Terrazzamenti" (half of the slope) and the "Sentiero Valtellina" (in the valley floor) are fragmented and lack sufficient infrastructure, making it difficult for tourists to enjoy a seamless experience. The region's limited transport infrastructure reduces its overall accessibility, hindering tourism development and economic growth. To address this, there is a pressing need for modern transport solutions.

Valtellina's natural environment faces significant risks due to land abandonment, soil erosion, and the region's continued reliance on fossil fuels for energy. The abandonment of traditional agricultural terraces in areas like Valmalenco and Val Gerola has exacerbated issues of soil erosion and hydrogeological instability, making these areas more vulnerable to landslides and flooding. This neglect of agricultural land not only impacts the landscape but also threatens the region's ecological balance. The ongoing environmental degradation caused by the abandonment of agricultural lands poses a serious threat to Valtellina's long-term sustainability and ecological balance undermining the crucial base for the development of a sustainable tourism.

Also, Valtellina's rich cultural and historical heritage remains underutilized, and many significant sites are deteriorating due to a lack of preservation efforts. Several abandoned villages, such as Cevo (Val Masino), Scilironi (Valmalenco), are at risk of being lost. These areas hold great potential for cultural tourism but require urgent investments in restoration and adaptive reuse to preserve their historical value and attract visitors. Additionally, neglected complex, including the former Fossati cotton mill in Sondrio and the abandoned sanatorium in Prasomaso, are resources for redevelopment.

The Post-Covid Change

The lockdown experience has been a turning point which made urban population rediscover the physical proximity of the low-altitude mountains to metropolitan regions. Furthermore, the covid widely promoted remote/smart working and made people aware of the outdoor activity benefits and of the lifestyle health potentials. This cultural change, involving living, working, vacation schedules and locations, turned out for whole Valtellina in a reversal of the shrinking and tourist dynamics since 2020 (Figs. 5, 6, 7, 8, 9, 10, and 11). The phenomenon is confirmed also on the base of data related to mobile connections: the overall presence of mobile users in

[16] Data: Chamber of Commerce of the Sondrio Province, annual report, 2023.

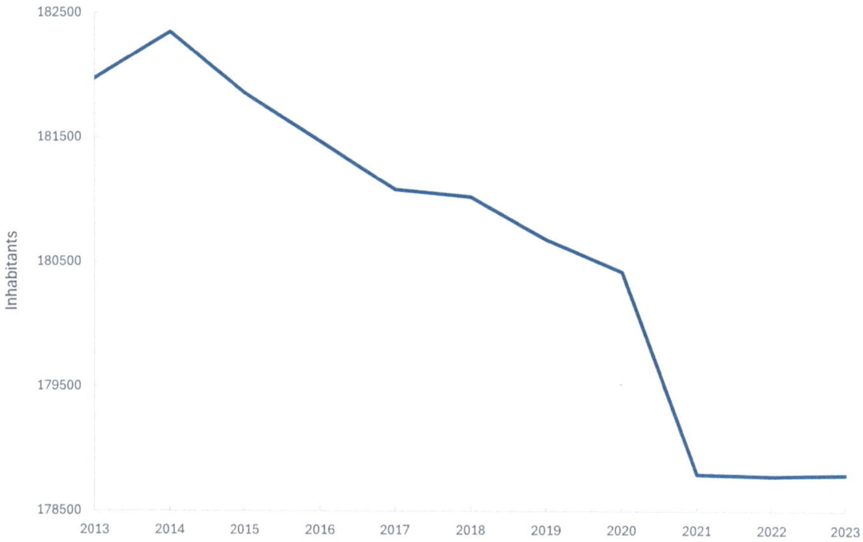

Fig. 5 Population trend from 2013 to 2023 in the Province of Sondrio. It is evident the after-covid change: population is no more shrinking as average on the whole Valtellina. Source: graphic elaboration by Deborah Paolini from ISTAT data

Valtellina has increased on average by 30% after covid (2020–2022).[17] Some re-population phenomena are observed also zooming on villages along the "Via dei Terrazzamenti" (Fig. 6).

The Proposal for a Smart Terrace Trail

The proposed project is focused on the main potential of the low-mid Valtellina, the so-called Route of Terraces, created in 2015 through a participatory process supported by a Lombardy Bank Foundation (Fondazione Cariplo). The careful selection, reconnection and arrangement of pre-existing historical traces, across the vineyard dry stone terraces, old villages and rural nuclei of the Rhaetian slope, allowed the creation of a continuous path for bike and pedestrians: 74 km long, the route goes up from the valley floor at Morbegno and runs all along the sunny half-slope "balcony" until going down to Tirano.

A bottom-up project which involved several stakeholders (first of all, the vineyard and cellar owners and managers), supported by the cultural district model (Della Torre, 2015): rooted in the local community and historical identity, integrated in the

[17] See Fig. 2 in the chapter Boniotti C., Pimpinella A., "Heritage and Mobile Radio: a Data-Driven Approach to Address Cultural Heritage Valorization".

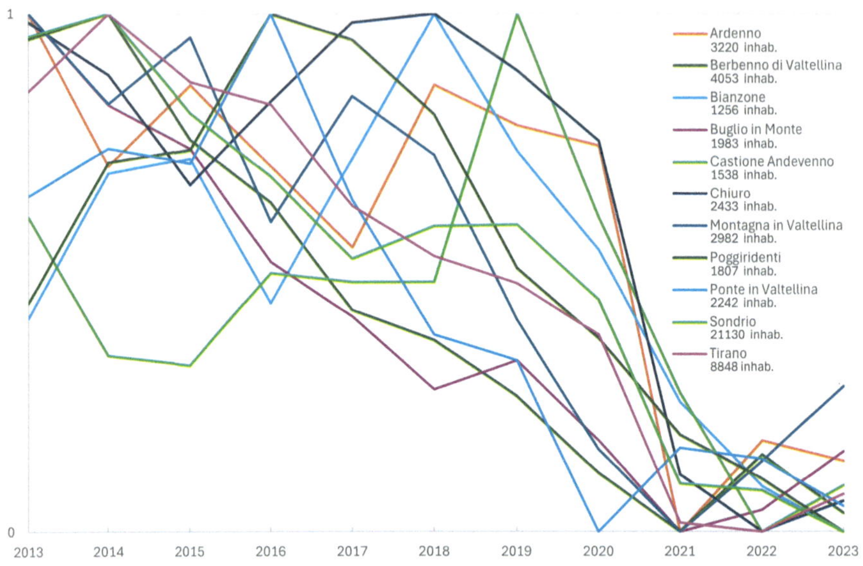

Fig. 6 Normalized population trend from 2013 to 2023 related to villages along the "Via dei Terrazzamenti". A post-covid inversion of the depopulation phenomenon can be noticed for several villages. Source: graphic elaboration by Deborah Paolini from ISTAT data

system of cultural and built heritage of the low-mid Valtellina.[18] Recently innovative farmhouses supporting eco-bike tourism are built along the Route:[19] we could consider them as spin-off of the Route of Terraces intervention.

Now smart/remote working and e-learning suggest a further evolution of the terrace route, providing the tech tool for a quality jump to become an innovative regeneration catalyst. The smart evolution of the Terrace Route envisages itself as strategic project for the regeneration of the whole low-mid Valtellina: we would rename it the "Smart Terrace Trail" indicating a linear platform combining working, education and vacation for innovative lifestyles (Fig. 12). Continuous digital accessibility along the path, plug-in spots, a sequence of coworking centers in each of the villages aligned on the trail: the Smart Terrace Trail will become not just a historical landscape experience and an outdoor physical activity but also a creative path which could host on-the-way brainstorming, professional strategic advice, art and culture production and events, team building programs. In synthesis, a prototype route for *smartwalking*. Of course, the performance of the path for experience tourism will be enhanced (food and culture) as well as for sport activities (fast walking, mountain bike, country running competitions) and under the therapeutic and wellbeing perspective, both physical and mental. It also makes treasure of a contemporary

[18] See chapter Boniotti C., Pimpinella A., "Heritage and Mobile Radio: a Data-Driven Approach to Address Cultural Heritage Valorization".

[19] See, for example, the eco-bike-agri hotels *Curt de Clement* in Tirano, *Contrada Beltramelli* in Villa di Tirano, *Olmi* in Sondrio, *Ca' Rossa* in Montagna.

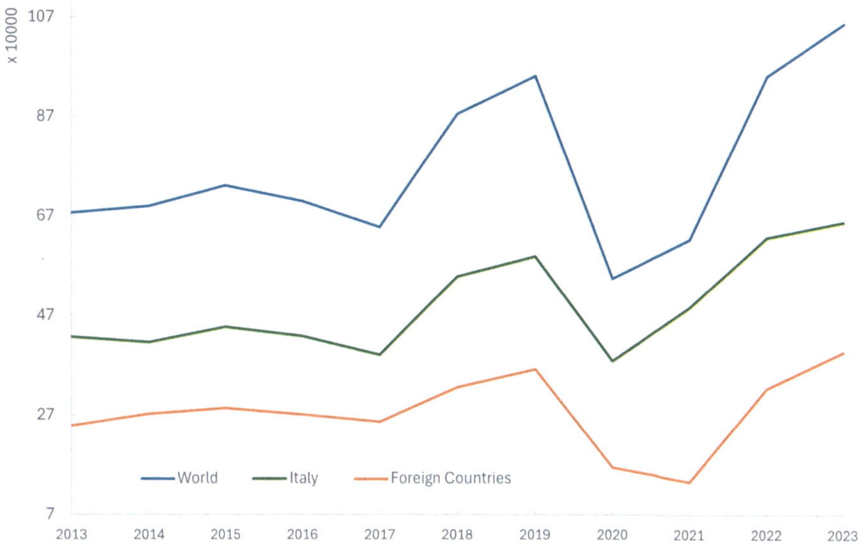

Fig. 7 Tourist arrivals from 2013 to 2023 in the Province of Sondrio. An intense post-covid recovery of tourist arrivals and an overshoot of pre-pandemic levels can be noticed. On average, Italian tourist arrivals nearly double international tourist arrivals. Source: graphic elaboration by Deborah Paolini from ISTAT data

dynamic reinterpretation of the Rockefeller research resort concept (ref. the nearby Rockefeller resort in Bellagio and other Lake Como research complexes, as Villa Vigoni, dating back to the '50 reusing XIX century villas).

Sensors along the path, connected with drones and satellites, will help monitoring the people flow for tourism and regional marketing purposes but first of all for weather forecast and environmental temperature in order to be able to adapt in real time route programs; also water derivations could be monitored in order to keep the micro landslides and in general the hydrogeological risk under control (Mayr et al., 2022). Climate change is introducing modifications in mountain agriculture: recently the ideal altitude for vineyards is raising, so lowest vineyards are substituted by olives trees (a news for Valtellina) and vineyards could extend their border above, reoccupying old domains once cultivated even if steep but then abandoned because of depopulation.

The new version of the Terrace Route is conceived as an integrated smart strategic spine to support a wider, multitask regeneration process.

Along the path, abandoned rural nuclei[20] could be regenerated to host small-scale co-working and co-studying spaces, meeting and workshop facilities, restaurants and coffees, art and artisan atelier and showrooms; all these facilities will be both

[20] E.g. regeneration resources are *La Curta*, an abandoned village below Teglio, and other dismissed farms or small rural factories on the Rhaetian slope.

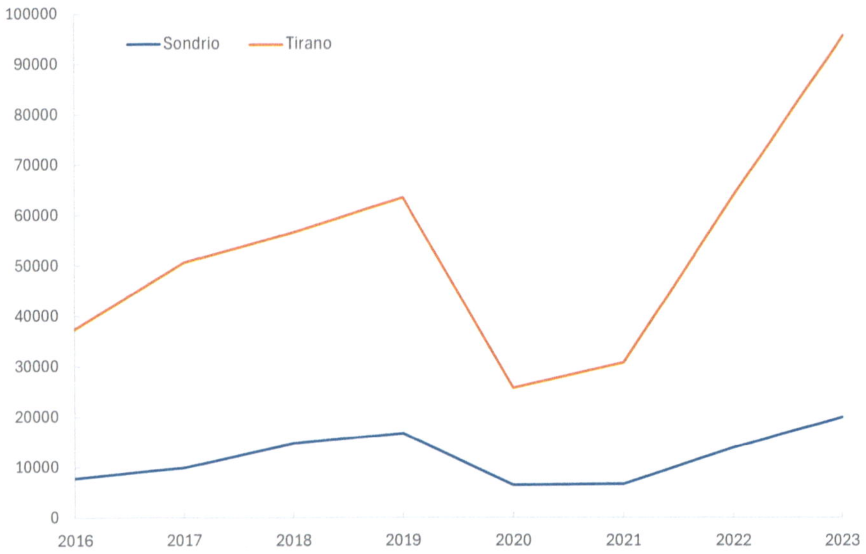

Fig. 8 Tourism evolution in the small towns of Sondrio and Tirano (tourist arrivals 2016–23). The tourism starts increasing few years before the covid and is re-booted by it. Source: graphic elaboration by Deborah Paolini from ABIT Osservatorio Locale del Turismo for 2016–18 data and from Polis Lombardia for 2019–23 data

indoor and outdoor. Vista points,[21] will be provided by plug-in availability and designed as outdoor creative spaces.[22]

On this platform both semi-abandoned upstream villages to be regenerated and downstream logistic bases will be grasped. Cores in altitude could be reused as corporate retreat (first of all, the former sanatorium of Prasomaso) or hermitage for digital detox or for rural bio-tech activities (the "alpeggi" above Boirolo and Pian dei Cavalli); the main anchors of this spine to the valley floor (Morbegno, Sondrio, Tirano) could host super high-tech co-working facilities integrated with e-learning centers, near the train station.

So, the concept is intertwining a horizontal linear model of digital facilities making the Terrace Route a prototype of super-equipped Smart Walking path (supported by built heritage regeneration) and a vertical transect model combining logistic bases in the bottom of the valley in coincidence with main train stations, smart creative facilities at the vineyard level half of the slope and digital detox above (Fig. 13). Drone services and intermobility services will complete the model, allowing tours

[21] E.g. smart vista points could be created regenerating the abandoned hut over Ponte, the existing "belvedere" over Villa di Tirano, the arrival building of the rural cableway inside the *La Gatta* vineyards, the *Calvario* churchyard and the panoramic bend overlooking the *Loreto Sanctuary* in Tresivio, etc.

[22] Cf. vista points funded by the Scenic Route Department of Norway along the *National Tourist Route*, designed by contemporary landscape architects.

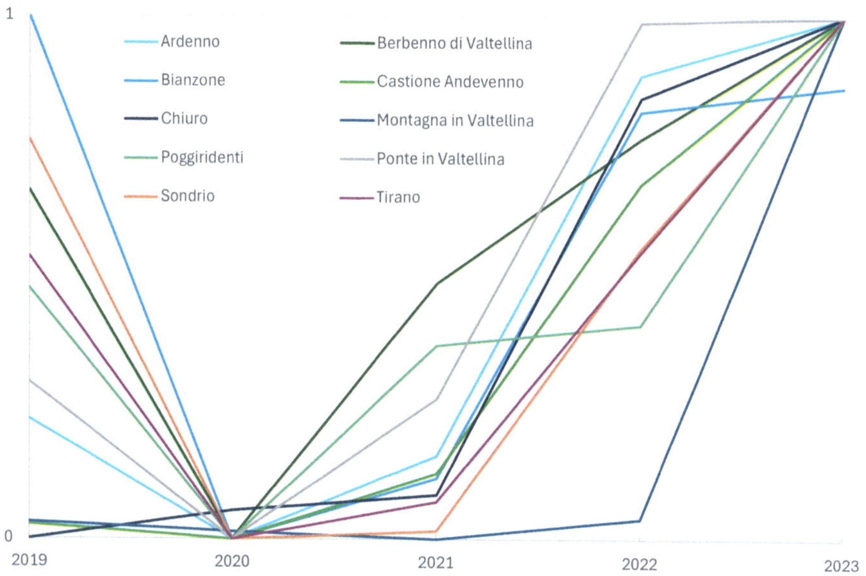

Fig. 9 Normalized tourist arrivals from 2019 to 2023 for several municipalities along the "Via dei Terrazzamenti". An overshoot of pre-pandemic levels can be noticed. Source: graphic elaboration by Deborah Paolini from Polis Lombardia data

served by various mobility ways (e-shuttles, train re-programmes, e-bike, walking/ biking).

The need to make a system is evident between Morbegno, Sondrio and Tirano in order to provide gateways and anchors to the villages of the "Via dei Terrazzamenti".

The highest capacity for regeneration belongs to the transect of Tresivio with the former sanatory buildings of Prasomaso.

The transect Sondrio-Valmalenco shows an on-going regeneration process both inside the city center (transformation of the main square, the proposal for a contemporary art museum in a former bank building, the new cultural/events pole at Castel Masegra) and above, at the entrance of Valmalenco, where a new pedestrian bridge overcomes the stream gorge providing continuity to the Terrace Route; further above, the abandoned village of Scilironi is calling for regeneration.

Crucial is the transect Teglio/Tirano/L'Aprica which has the potential of connecting Valtellina with Val Camonica and beyond, towards Venice; on the opposite side, this transect connects Valtellina with Valposchiavo towards S. Moritz.

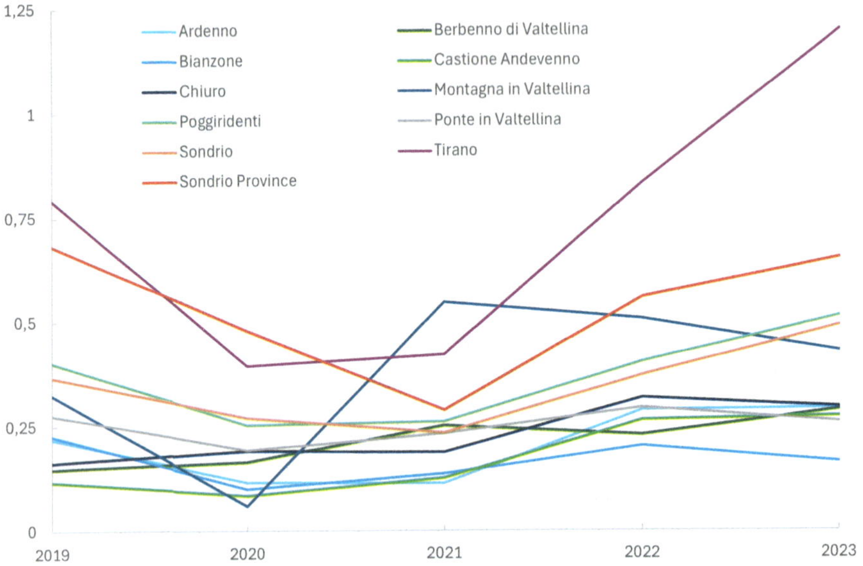

Fig. 10 International vs. Italian tourist arrivals ratio from 2019 to 2023 for the Province of Sondrio and municipalities along "Via dei Terrazzamenti". Despite Italian tourism prevalence, international tourism is present and shows a growing trend, especially in Tirano. Source: graphic elaboration by Deborah Paolini from Polis Lombardia data

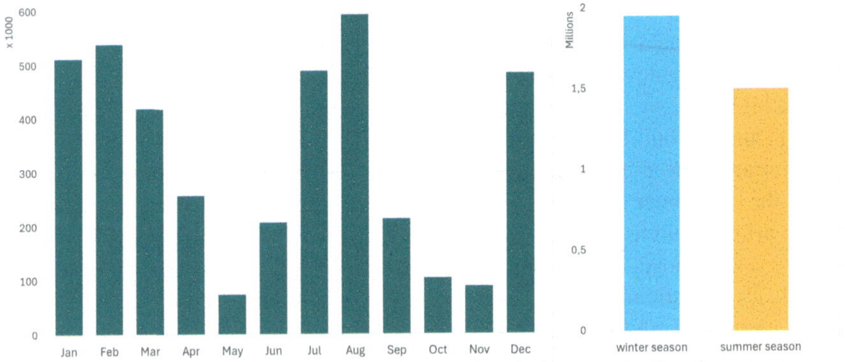

Fig. 11 Tourism distribution during the year, Province of Sondrio 2023. Winter season is predominant but also summer season is significant, especially for the highest peak of the year which is in August. In order to get a year-around tourism presence balanced through different months, the low-mid valley experiential tourism potential could help filling the lack of tourism in the dead months of May and October/November. Source: graphic elaboration by Deborah Paolini from Polis Lombardia data

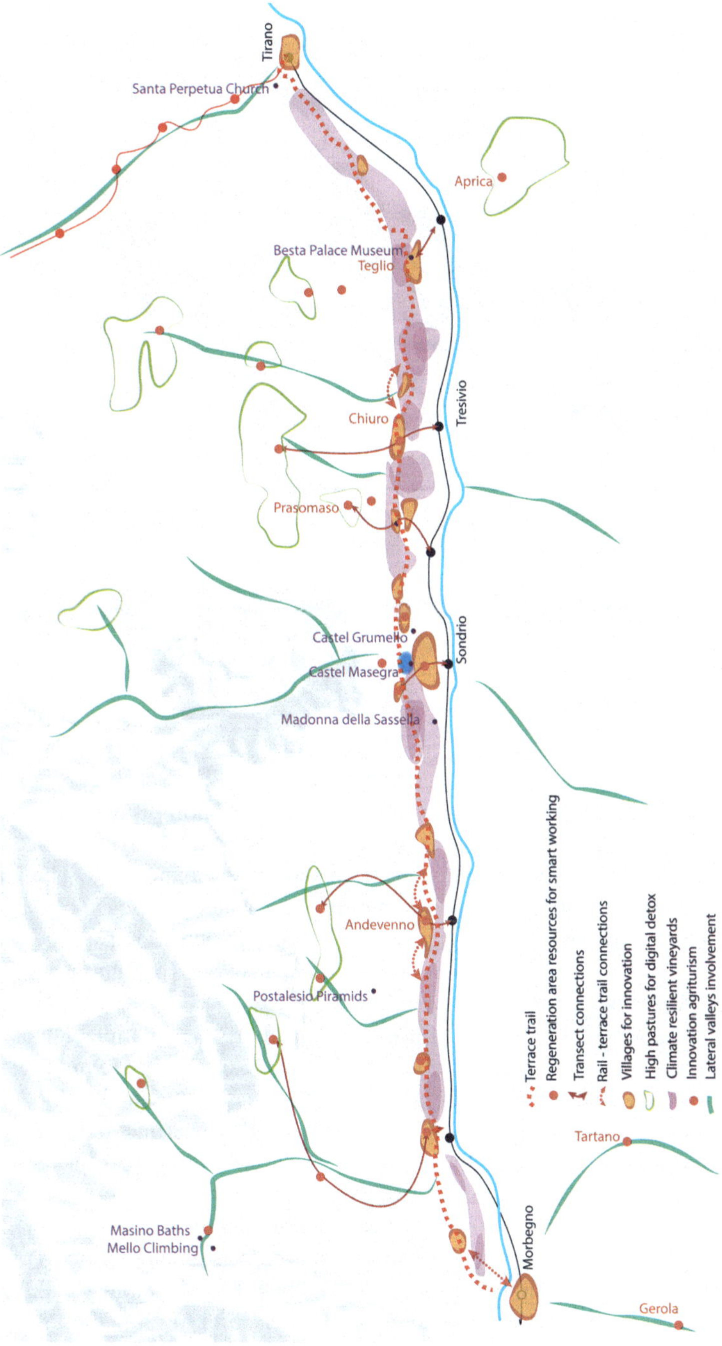

Fig. 12 The proposed Smart Terrace Trail Concept Plan. Source: image by Felix Günther

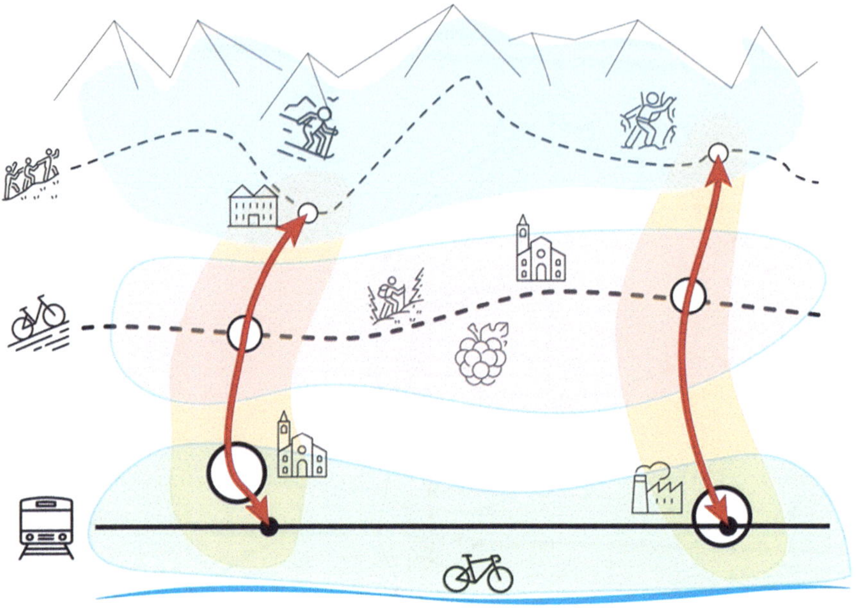

Fig. 13 The Smart Terrace Trail model combines a smartwalking linear platform with logistic/service transects between valley floor and ridge systems. Source: image by Felix Günther

A Multiscale Approach

The Smart Terrace Trail project is rooted in a multiple-scale approach: it is a prototype of a system of paths connecting historic centers and hiking itineraries, urban open spaces with high routes across excursion landscapes. The territorial scale is strategic for the updating concept of open space and greenway network. The long-term scenario foresees also the regeneration of the degradated landscape of the valley floor along the Adda river greenway, in order to restore perceiving space for the survived built heritage and to mend the ecological network (preventing climate change and hydrogeological risks at the micro-scale).

The evident potential of the Trail for experience tourism is complementary to upper slopes vocation for sustainable sports (such as excursions, climbing, ski mountaineering and snowshoes). The so called "alternative mountain" model (a mountain which rejects ski-lifts and cable-ways), recently experimented in the village of Montespluga, at the Valtellina Northern edge, is suitable also for the high landscapes of the low-mid Valtellina. Existing transversal rural routes could connect the Terraced Trail with the higher ridge paths which follow the valley watershed.

Zooming out at the macro-scale allows us to interpret the Smart Terrace Trail as central section of a contemporary "wine path" stretching itself all along the Pre-Alp arc: starting in Piemonte and continuing across Franciacorta towards the Prosecco Hills (recently Unesco Heritage) and Friuli vineyards.

Actually, digital transition supports a new vision at a wider scale not only as thematic (wine) perspective but as integrated planning. Given the possibility of remote/smart working and e-learning above described, strategies could be developed for a resilient polycentrism, at regional and cross-border scales, which foresees the low-mid Valtellina as an emerging pole. Digital accessibility could reposition this territory at the center of the Alps despite its difficult physical accessibility (even beyond the hypothesis of a new gallery under the Mortirolo mountain connecting Tirano with Val Camonica and Brescia urban region). Making treasure of being not far from Milan and Zurich metropolitan areas and in between global tourism destinations such as Como Lake and S. Moritz, low-mid Valtellina could also develop an offer for new lifestyles and for experiential/sustainable tourism in perfect complementarity with the urban luxury mountain model of S. Moritz and the grand tour model of the Como Lake.

The Smart Terrace Trail, indeed, is called to become the strategic central section of the "slow" transalpine corridor connecting Milan to Zurich through Como Lake, Valtellina, Engadina.[23] The multi-scale integration of sustainable mobility ways with services, featuring the proposed project, is inspired by the Swiss strategies for mountain settlements which consider a value that people born in a mountain village will continue to live and work there.[24] The project is developed through a site-specific and digital-innovative approach which gives value to the uniqueness of this productive landscape.

Zooming in to Regeneration Areas as Key Factor

Regeneration areas are a pivotal component in the advancement of the region. In the context of tourism development in a shrinking region, it is of the utmost importance to prioritize the concept of internal development and the zero soil use agenda. A case in point is the failure to secure the UNESCO label for the terraced vineyards, which was due to the insufficient spatial qualities of the buffer zone around the heritage sites, resulting from urban sprawl. It is therefore imperative to identify a means of ensuring that future economic development does not compromise spatial quality through the utilization of new land, as has occurred in recent years in Valtellina. Instead, efforts should be directed towards the respectful regeneration of abandoned industrial areas and deserted villages. Our research has demonstrated that the following areas and villages warrant particular attention regarding the development of Valtellina. In this context, it is of paramount importance to give priority to the respectful regeneration of abandoned sites over the use of new land for economic development in Valtellina.

---

[23] See chapter Fossa G., Günther F., Rollandi A., "A Sport and Heritage Corridor for the Valtellina Olympics".

[24] Cf. Günther F., Grison Strategic Plan 2014.

Several dismissed industrial areas in Valtellina hold significant potential for regeneration and integration into the region's tourism development strategy. One of these is the former Fossati cotton mill in Sondrio, an area of 30,000 sqm, laid out between the terrace trail and the Castello Masegra dominating the city. Plans for its development focus on transforming the site into a multifunctional hub that includes coworking spaces, tourist accommodations, and cultural facilities. This project aims to cater to digital nomads and long-term tourists, while preserving the region's industrial heritage.

Another key site is the Sanatorium of Prasomaso located in Tresivio, near Sondrio, a huge, abandoned complex built at the beginning of the XX century to host tuberculotic patients in a forest area of 60,000 sqm perfectly exposed to South, at the altitude of 1200 m asl: the sanatorium was abandoned in the '70s, a typical dismission phenomena of the entire Alps region. The preservation of the interesting liberty architecture represents a heritage value for the regeneration potential as corporate retreat: the international real estate market is sensitive to opportunities for this kind of function in iconic historical buildings located in quite remote inspiring places but not far from international airports.[25] It has the potential to become a wellness center and corporate retreat, accommodating for both tourists and professionals, emphasizing on wellness and eco-tourism.

In Teglio, the former AEM colony, a dismissed structure of about 37,000 sqm, can be converted into a tourism hub, featuring lodging, coworking spaces, and cultural activities attracting locals, tourists and digital nomads by providing sustainable accommodations in a historical setting, further enhancing Teglio's appeal. Meanwhile, in Tirano, the abandoned railway yards South of the main station offer almost 30,000 sqm as node for a larger regeneration project of the whole city center after the opening of the Tirano road bypass. Furthermore, different smaller regeneration sites as the former Mottana factory, an area of 3000 sqm, offer surfaces to be integrated.

Several abandoned villages in Valtellina show potential for regeneration and development into unique tourist destinations. In Val Masino, the village of Cevo, located at mid-altitude, could be set to have its abandoned homes restored and repurposed as part of a "hotel diffuso" (scattered hotel), also considering the potential of a geothermic energy community. This approach would integrate touristic accommodations with the regeneration of local heritage, supporting "slow" and "eco" tourism while revitalizing buildings in an area of significant cultural value. In synergy, the above historical Masino Baths are calling for regeneration. In Val di Mello, abandoned properties offer the possibility to create reception facilities for the rock climber sustainable tourism attracted by the local granite cliffs. In Valmalenco, the abandoned villages of Scilironi and Bedoglio could be slated for restoration. In Val Tartano, the village of Tartano also offers potential to be regenerated as a site for agritourism and wellness tourism as could the semi-abandoned area

[25] See for example the Novartis corporate retreat on the Zug lake in Risch (CH).

of La Curta in Bianzone reviving the village's agricultural traditions as a scenic stop along the "Via dei Terrazzamenti".

These projects would aim to blend historical preservation with contemporary visitor engagement using modern technologies like augmented reality to create immersive historical experiences, attracting cultural tourists seeking to explore Valtellina's rich history in a novel way. The scattered hotel, where hotel rooms or suites are spread across different buildings in a village, allows guests to experience living in local homes or historical buildings while still benefiting from hotel services, providing tourists with an eco-friendly and immersive experience in nature. This development aligns with nature-based and sustainable tourism, aiming to revive the area while preserving its natural beauty.

A Sustainable Mobility for Tourists and Residents

The enhancement of internal connectivity within the Valtellina region represents a pivotal element in the advancement of sustainable mobility and the optimization of the sport&heritage transalpine corridor.[26] A number of significant infrastructure developments have been proposed with the objective of reducing reliance on private vehicles and facilitating tourist access to the region's cultural and natural attractions. One such initiative is the introduction of electric shuttle services along valley transects, departing from major hubs such as Tirano, Sondrio, and Morbegno along the railway line. Key to all these approaches is a railway as backbone of the collective mobility system with frequent offers (every 30′) and comfortable and equipped high quality rolling stock (plug-in, WIFI, bike-transport for all classes etc.).

The linear trail offers a distinctive opportunity for walkers to gain a comprehensive understanding of historic landscapes, with the ability to continually gather new impressions along a continuous route. Such way facilitates a more profound engagement with historical or cultural landmarks, as the emphasis is on a linear trajectory that frequently traces the routes previously used for trade or pilgrimages. One significant benefit is that hikers are not required to retrace their steps, which fosters a sense of exploration and enables uninterrupted forward movement. However, the one-way nature of such routes necessitates the availability of a functional and reliable public transportation system with convenient services that provide satisfactory amenities during any waiting periods. In contrast, circular routes, such as the "Strada del Vino", offer the possibility of returning to the starting point, which is often a car park opposed to the aims of ecological tourism.

These services will connect key locations and villages along important routes, including the "Via dei Terrazzamenti". The "Sentiero Valtellina", along with other popular cycling paths, should be developed further, as should routes leading to the

---

[26] See chapter Fossa G., Günther F., Rollandi A., "A Sport and Heritage Corridor for the Valtellina Olympics".

high-altitude villages and sanatoriums. The aim is to minimize car usage and promote eco-friendly tourism, allowing visitors and the workforce to move seamlessly between sites without contributing to environmental degradation. Furthermore, the potential for the development of funiculars should be considered, connecting the various altitudes from the valley floor to the "Via dei Terrazzamenti" and higher areas with significantly reduced travel times, offering an efficient alternative to car travel.

The expansion of cycling routes is key to improve the overall accessibility with an integrated collective mobility system, focusing on the extension and integration of the "Sentiero Valtellina" (along the Adda River) and the Ciclabile Valchiavenna with other offers as the "Via dei Terrazzamenti" and the "Strada del Vino". These routes will be enhanced with better signage, rest areas and infrastructure to connect cycling paths with cultural sites and local towns. This initiative aims to promote slow tourism, walking and cycling as a primary mode of transport for tourists, encouraging a sustainable and immersive exploration of the valley. To further improve pedestrian connectivity, it is essential to implement improvements to the final sections of the "Via dei Terrazzamenti" where the route still uses busy roadways. In particular, the section East of Chiuro could be mitigated with the construction of a new pedestrian bridge towards Ponte in Valtellina. Furthermore, the western village exit of Andevenno, which currently shares the road with motor traffic on a longer, blind stretch, could be improved. Finally, the village exit of Postalesio would benefit from the provision of alternative routes or, at the very least, a safe pedestrian and cycle path parallel to the road.

Rooting Innovative Tourism in Local Communities

In conclusion, this project underscores the potential for transformative change of shrinking low-altitude mountain regions (like the middle Valtellina) through a comprehensive, experience-centered approach to tourism development, deeply rooted in local communities. The integration of cultural heritage, local gastronomy/agriculture, sustainable mobility with digital accessibility and coworking spaces exemplifies how experiential tourism can serve as a catalyst for economic revitalization, social resilience, and environmental sustainability. This model addresses the complex challenges faced by rural, depopulating areas, such as demographic decline, economic stagnation, and the underutilization of cultural assets, by fostering a year-round tourist economy that aligns with local traditions and site-specific cultural/ecological values. Furthermore, the Smart Terrace Trail's emphasis on collective mobility demonstrates a dedication to reducing the environmental impact of tourism while enhancing accessibility and connectivity. Such infrastructure, when coupled with digital facilities and engaging visitor experiences, ensures that the destination appeals not only to traditional tourists but also to digital nomads and remote workers, promoting Valtellina as a dynamic and adaptable landscape.

Acknowledgements Authors would like to express special thanks to Dario Foppoli (partner and founder of Foppoli Moretta Associated Engineering, Tirano) and to Marco Scaramellini (Mayor of the City of Sondrio).

References

Dania, V., & Gazzerro, L. (Eds.). (2023). *Fondazioni e beni ecclesiastici di interesse culturale. Sfide, esperienze, strumenti*. Il Mulino.

Della Torre, S. (2015). Shaping tools for built heritage conservation: From architectural design to program and management. Learning from "Distretti culturali". In K. Van Balen & A. Vandesande (Eds.), *Community involvement in heritage* (pp. 93–101). Garant.

Foppoli, D. (2019). Le costruzioni in pietra a secco nel paesaggio culturale valtellinese. *Notiziario Istituto Archeologico Valtellinese, 17*, 123–146.

Foppoli, D., Menghini, G., & Zanolari, E. (2014). Preventive conservation: An opportunity of cooperation in the hearth of the Alps. In *Proceedings of the international conference preventive and planned conservation, vol. 1, Monza, Mantova, 5–9 May 2014* (pp. 85–95). Nardini Editore.

Mayr, A., Scaioni, M., et al. (Eds.). (2022). *Sensing mountains*. Innsbruck University Press.

Moioli, R. (2023). *La conservazione preventiva e programmata: una strategia per il futuro: premesse, esiti e prospettive degli interventi di Fondazione Cariplo sul territorio*. Nardini.

Schurch, D. (2006). *Nomadismo cognitivo. Ingegneria dello sviluppo regionale*. FrancoAngeli.

Varotto, M. (2020). *Montagne di mezzo. Una nuova geografia*. Einaudi.

Stakeholder Perspectives for Regenerating Inner Territories

Cristina Boniotti and Giovanna Fossa

Research Colloquia

A continuous dialogue and discussion with experts, stakeholders, local municipalities, and multi-scale institutions and enterprises supported the research path by addressing the interpretation of places and regions, thus allowing the proposed strategies to be rooted in sustainable and feasible frameworks. This process is represented in the following selection and synthesis of the many colloquia that accompanied our research. From a methodological perspective, it is evident the research challenge on the language issue which requires a cultural mediation both for fine-tuning the discussion among various disciplinary fields and for intertwining the academic research with the life-and-work experience in places.

Through the various professional, research, and government experiences and through the different testimonials of stories lived in places, the whole range of tools present today in public agendas and innovative experimentations, aimed at revitalizing depopulating villages, appears. These approaches can be interpreted as innovative combinations: contemporary art & historical heritage, traditional crafts & contemporary design, zero-km production & e-commerce, organic & tech-agriculture, active mobility & health, digital hyperconnectivity & detox, tech innovation & preservation, remote working & community experience.

Even with their different nuances, the perspectives of experts and stakeholders converge on the fundamental issue for the various tools to bear fruit over time, considering local communities and their rootedness as assets, whether that be native or chosen, and investing in their culture and education. The vision that emerges is

C. Boniotti (✉) · G. Fossa
Department of Architecture, Built Environment and Construction Engineering, Politecnico di Milano, Milano, Italy
e-mail: cristina.boniotti@polimi.it; giovanna.fossa@polimi.it

G. Fossa et al. (eds.), *Small-Town Renaissance*, Research for Development,
https://doi.org/10.1007/978-3-031-89786-3_9

connoted by strategic binomials such as site & culture-specificity, place branding & community self-image, fresh-eye research & lived experience. What is common to all of the above is a special focus on the global trend to move beyond mass tourism that consumes the landscape toward tourism that pursues the experience of an authentic, lively, and productive territory.

We Have Everything, the Rest Is Missing: A Platform for Change

Synthesis by the authors of the research colloquia with Andrea Bartoli, Founder of Farm Cultural Park, Favara (Sicily).

Notary Andrea Bartoli and lawyer Florinda Saieva, an art-collecting couple with a great passion for contemporary culture and heeding the dream of improving small portions of the world, founded Farm Cultural Park in 2010. The parents of two girls, they longed to do something to reverse the dereliction and marginality of their home town of Favara, an ancient village occurring less than 10 km from Agrigento (population: about 30,000) that had been thriving since pre-history but in recent times suffered a severe depopulation trend due to a lack of opportunities aimed at young people and families alike. The couple envisioned the opportunity to inverse the trend in place and halt the overall structural, social, and economic decay of the city by resorting to art and architecture. Instead of waiting for someone to come to their aid and change their city, they decided to devote their whole lives to some ambitious transformation process. Farm Cultural Park is now a genuine workshop and hothouse of social innovation, a space inside of which a community of both locals and creative talents work on problems and intervention strategies, seeking to make the most of their resources to reuse, regenerate, reinterpret, revitalize, and cultivate what's available to them. Their main objective is to beget cultural transformations in urban places to bring to light the possibilities of human creativity while targeting today's challenges and opportunities. For a town that until recently had been unknown even to many a Sicilian, the regeneration experienced by Favara amounts to nothing short of a small miracle. In a Nation of abandoned villages, Favara is beating the downtrend, and using its past as a foundation to build a future rooted in the fascinating energy of art and culture.

The process of propagation of new practices initiated in Favara is a surprising one. Farm also started an additional cultural hotspot in the Sicilian village of Mazzarino (population: 10,000) in 2022 and is now opening a third one in the very heart of Palermo, thus reaching a total of 15,000 sqm of regenerated areas. The Palermo spot is introducing a multi-scale dialogue between villages and cities which will hopefully benefit the resilience of the Farm Cultural Park system as a whole—the small town of Favara will nonetheless continue to be their flagship center and process driver.

Each location boasts its own array of experimental research projects, laboratories, and creative productions in the realms of art, education, environment, and placemaking.

Farm thus came to be as a family initiative, in line with the fertile tradition of Italian family businesses, yet in the shape of a non-profit foundation focused on the transformative action of culture and aimed at the involvement of the local community and a vast network of relationships and instances of friendship. The end result is a community variable yet well-rooted, either by birth or by choice, in the places to be regenerated (mostly as volunteers, some as workers). Professors from universities such as the Politecnico di Milano are also involved in this network, and over a hundred graduation dissertation thesis have dealt with the iconic case of Favara.

Farm captures an outspoken demand for innovative communities. The disruptive force of an organization like Farm is the informal scheme that first allows people in the community to question their vision of themselves and their future with respect to a place, and then to bring actors together, whilst adding to that an experiential dimension of international comparison. Important artists and architects from abroad are involved in the regeneration not only of artifacts but of the community in itself; Farm also takes care of the crucial aspect of continuity of international relationships over time. The role of world-wide networking is a strong one: as to the latter objective, Farm acts in cooperation with the Moleskine Foundation, which promotes international networks of socio-cultural organizations, even small-scale ones, by supporting them with a mini-token of a few thousand euros which sometimes, as is the case in some Central African contexts, may allow for some quantum leaps in intervention quality.

International comparisons (e.g., the biennial event "Countless Cities" telling of city stories on a worldwide scale) are integrated with projects specifically aimed at Sicily's inner territories. To this effect, we ought to mention the collaboration in place with the Municipality of Aragona aimed at the opening of an unfinished auditorium currently being returned to the community through the work of architects and designers. The 2021–2026 project going by the name "We have everything, the rest is missing" (a quote from Sicilian actor and writer Pino Caruso) is an exhibition spread in space and time devoted to Sicilian artists and aimed at experimenting the role of art throughout a path of social transformation and regeneration. Of note here is also "Marea", a transformative action being exhibited as a work of art: a community movement of Sicilians by birth or choice who fund creative regeneration projects made by Sicilians in their home land.

In the 3 years between 2015 and 2018, Favara's accommodation potential grew from 6 to 600 beds. In spite of that, the suspension of hotel activity begotten by the pandemic-induced lockdown has made it clear that private initiative does not suffice by any means—what it truly takes is an involvement of institutions to enable hamlets and cities to weather the ups and downs of urban fads that lead them to cyclically grow and decline in attractiveness. Supportive public policies are especially needed to ensure continuity beyond the founders' initiative. Farm is financed by the family of the founders who devote 70% of their working time to it by consulting services serving Municipalities, companies and schools, calls from the Ministry of

Culture (Creative Lab), and a few sponsors; structural supportive actions would be needed, such as the ministerial recognition of Farm as a place of training for public administrations and local communities, starting with children. In the years following the Pandemic, the mission of working more and more with institutions to innovate them from the inside through art and creativity has become a clear one, thus leading such institutions to fearlessly experiment with different policies. The one of institutions in Sicily as well as throughout the south of Italy is a big problem: 90% of Municipalities are in a plight of economic default; what's more, the issue is not even part of anyone's political agenda.

Experiments characterized by a cooperation between private initiatives and public institutions are carried out by Farm especially in the fields of schooling and migrant reception. Recently, a sign of hope has opened up in this perspective thanks to the Prefecture's support for Farm's initiative related to the migrant hotspot of Porto Empedocle, where Farm created the "Transition for" group aimed at helping migrants manage their relations with the Prefect's office. Farm's commitment to education dates back to 2006 with the opening of the first Children's School of Architecture in Favara: there are now 29 such schools in various Italian cities which enjoy the support of major architectural firms such as Cucinella in Milan and Femia in Genoa. The school provides once-weekly afternoon classes to train children in vision and creative imagination on an annual theme shared among all schools pertaining to the system, including field trips and in-person meetings with renowned architects in European capitals. Indeed, the regeneration of local communities as a foundational strategy for revitalizing declining villages and towns starts with the training of children.

Digital Infrastructures Are Spreading But Usage Policies Are Late

Synthesis by the authors of the research colloquia with Antonio Capone, Professor of Telecommunications, Politecnico di Milano, Department of Architecture, Department of Electronics, Information and Bioengineering.

Italian and European public policies remark the importance of telecommunications as a gateway to digital transition. The main references for the European Union on this matter include European Digital commissioner Breton's White Paper (European Commission, 2024), the report on the market evolution by Letta (2024), and the report on European digital competitiveness drafted by Draghi (2024): all of them have placed telecommunications (TLC) at the center of digital development in the EU, and have defined strategic guidelines subsequently adopted by national governments in various ways.

Italy once suffered from a significant lag in telecommunications development due to the inertia of historical national monopolies. With the gradual opening to private entities, the digital access scenario has been characterized by a relatively competitive environment in areas with high infrastructure penetration combined with public intervention digital upgrading plans throughout underserved areas. Remarkably, Italy has been significantly closing its digital infrastructure gap over the last decade. Digital coverage now extends to almost the entire national territory, except for approximately 500,000 "scattered homes" located either in inland areas or some metropolitan outskirts, as well as the so-called "grey areas" typically served by weaker connections. Both issues have been clearly recognized, and will be addressed in the years to come. In fact, national plans foresee the extension of fiber optics along the entire road network, including secondary roads, to reach every address in the country by 2026 (the Italian BUL plan for Ultra-Broadband).

In this context, non-terrestrial (supernational) infrastructure based on satellite connectivity for broadband access, an item currently on the European political agenda, would enable coverage beyond physical addresses, where the fiber-grid is typically impossible to reach (e.g., remote areas and open seas). This would also be useful in case of catastrophic events, for navigation or military needs, and to reach temporary settlements such as alpine pastures or shelters, as an alternative to the FWA (Fixed Wireless Access) technology. In fact, FWA and satellite are already being offered as options for broadband access by several local telecommunications operators. Unlike fiber subscriptions, which are mandatorily annual, such operators offer temporary subscriptions (monthly plans), thus making seasonal or short-term packages more meaningful than annual subscriptions in some internal areas, even in such cases where fiber is actually available. Indeed, 5G-powered access is not necessary for remote work, as 4G-based FWA or satellite-based access are well sufficient in terms of performance.

From European public policies, tensions emerge between geopolitical and local interests as far as telecommunications are concerned. There is a lack of alignment with strategies for internal areas, a critical issue especially on Italian soil in light, also, of the tourism value of such regions. National telecommunications policies require more participatory development, closer co-operation with local communities, and a special focus on developing local site-specific services to support smart/remote working and address territorial imbalances, to wit, the main determinants of commuting and depopulation.

A sizeable growth potential exists in this area considering, also, the potential benefit that the revitalization of internal areas could bring on global competitiveness. In fact, the productivity gap between Europe and North America is almost exclusively due to digital companies (i.e., not manufacturing ones). Despite the growth of the Europewide telecommunications infrastructure, including Italy (and even though the 5G network is a laggard throughout Europe compared to North America and China, with the latter fact nonetheless being of barely any interest to our effect), the EU digital ecosystem has not developed, leaving Europe colonized by American Big-Tech companies (as has been noted in Draghi's analysis).

In contrast, China's model is profoundly different, being extremely efficient and pervasive across the entire population: the analog-world has fully disappeared (as even beggars on the street request money via WeChat). In spite of the above, neither the USA nor China prioritize internal area development. In the former case, production is increasingly centralized, while in the latter case the growth of rural areas does not pose a strategic priority. However, whilst Europe and Italy have a spatial structure different from both the USA and China, there is no evidence that investment in internal areas will boost productivity. In recent years, a Europewide and Italy-wide trend of colonization on part of multinational corporations (generally American) has been observed. These companies hire workers for full-remote jobs, taking advantage of high professional skills at lower salaries (notably through direct hiring, unlike call centers where hiring is intermediated by specialized companies). This allows people to work from anywhere, presenting a significant opportunity for internal areas, though the added value created is taken out of Italy. This can be considered the most significant post-COVID change: we keep exporting labor while at the same time staying at home.

This further confirms the need to integrate internal area with telecommunications policies. The focus on services should be aimed not only at meeting individual user needs, but also at supporting local communities—or even establish them to begin with. Once again, the digital divide primarily emerges as a cultural issue.

Microidentities and the Promotion of Territorial Cultures

Synthesis by the authors of the research colloquia with Michele Da Pozzo, Director of the Natural Park of Ampezzo Dolomites.

Cortina d'Ampezzo is undergoing a depopulation trend by gentrification, i.e., young people from Cortina move to the downstream villages to be able to afford some housing. Cortina is in a condition of overtourism and therefore would not be in need of the promotion provided by the Olympic event (Milan-Cortina 2026) and the UNESCO declaration (Dolomites World Heritage Site 2009). Promotion, even if taking advantage of the latter opportunities, should rather be aimed at the lower altitude mountains currently at risk of abandonment and depopulation.

Both in the Alps/Pre-Alps and Apennines, overtourism and depopulation areas have a common denominator, to wit, a micro-identity: we ought to be aware of this aspect and recognize it as an Italian uniqueness.

The entire Alpine arc shows signs of contamination between the overall Italian civilization and the specific cultures of inland valleys characterized by Celtic, Longobard, Occitan, and Ladin influences; the same applies to the Apennines and island mountains, where Albanian, Norman, or Iberian influences can be perceived. The result is the presence of exceptional toponymic languages, religious cults, soil regimens, and ways of cultivating the land. The so-called Ampezzan Rules (Regole Ampezzane), which governed the collective, indivisible, and inalienable ownership of forests and pastures in the hands of local family communities, are an illustrious example of this, as they led to a millennium of sustainable land management.

The first valleys to lose these identitarian values were those close to large metropoles, while the greater isolation of other territories allowed them to maintain linguistic, ethnic, and other forms of endemisms, such as the Ladin language.

In general, side valleys have maintained these peculiarities more visibly than transit valleys and especially the latter's valley floors. Think of the valleys of the Adda, Adige, and Piave rivers, which were "colonized" by exogenous civilizations, filled with defensive works, and have become major communication routes, whilst in the meantime their side valleys have retained more peculiar identities.

This distinction also arises from a bioclimatic perspective: north–south valleys, namely those serving as major transalpine communication routes, are open to northerly winds (e.g., the Piave, Tagliamento, Adige, Ticino, Ossola, and Sesia valleys), while the east–west transverse valleys are entalpic (thermally isolated) valleys characterized by a peculiar continental climate consistent with hot summers and cold winters (e.g., the Valtellina, Val Venosta, Val Pusteria, Val di Susa, the valleys of Cuneo, and Valle d'Aosta), with this outspoken difference in climate begetting a wholly different agriculture and mountain civilization.

The east–west valleys boast lots of historic terracing on their sunny slopes for cultivation, with historic rural routes now recovered for tourism (the "Route of Terraces" among the vineyards of Valtellina, the "Tobacco Route" in Valsugana). In light of their being very dry valleys with very low rainfall, they are also historically provided with canalizations aimed at water optimization whose maintenance paths are being recovered for hiking (think of the "Vals", water pipes carrying water from the glaciers to the terraces, rediscovered in Val Venosta with the "Valveg" track created for their maintenance).

This valley orientation-dependent bioclimatic diversity is intertwined with a geological variety drawn at a macroscopic level by the "Insubric Tectonic Line" separating the calcareous from the crystalline pre-Alps and running through the Valtellina, the Orobian Alps, and the granitic-crystalline Bernina massif; this line crosses halfway up the Rhaetian slope in Valtellina and then runs through the Val di Sole, Val di Non, and Val Pusteria.

The above micro-identity is even more outspoken in mid-altitude mountains and pre-Alpine valleys at greater risk of abandonment, compared to mountains raising above 1000–1200 m and still having sufficient natural snowfall for winter sports.

The Italian Pre-Alpine arc from the Carnia region all the way to Liguria boasts a heritage and biodiversity value much greater than the inland valleys of the Alps as the former underwent earlier anthropization in the virtual absence of glaciations: therefore, they have a greater variety of crops and endemisms, and excel in biodiversity.

This is the great treasure of mid-altitude mountains.

We need a vision of sustainable development that enhances this richness of micro-identities rooted in history, supported by local community entrepreneurship that is aware of these values. The Olympics are not a solution, but rather a risk if they only result in procurement in the absence of cultural promotion of the territories. High-end tenders do not suffice; we need to know how to integrate the historical and identitarian values of places, also seizing the opportunities offered by remote working tools.

There is often a cultural barrier not so much between valleys (with ridges and watersheds separating them), but rather between valley floors and slopes: valley floors are urbanized and industrialized, even when it comes to crops (think about the apple orchards of Val Venosta), while historical cultivations monopolize the slopes to shape the landscape and provide hydrogeological protection (e.g., vineyards on the slopes in Valtellina compared to the degradation of the valley floor). A cultural barrier that in addition to cultivations is reflected in building types and settlement and tourism patterns.

Ease of access can also be a disadvantage to mid-altitude mountains: if a valley is too close to a metropolis, the odds of there being some form of cheap, "hit and run" weekend tourism will be greater. In contrast, isolated mid-altitude mountains bear some extra potential but also need to offer high-quality services (the historic villages of Vorarlberg, embedded as they are in cycle-pedestrian domains that can only be reached by shuttle may serve as an excellent example thereof, with another virtuous reference being that of the cores of Erto and Casso in Vajont recovered as a widespread village of stone houses to be rented out to mid-level foreign tourists, groups of friends and families, etc.).

Cultural heritage, including intangible heritage, is endangered by pervasive processes of globalization that spread, by way of example, New Age or Buddhist-inspired tourism practices, whether for physical and mental wellness or climbing, while territory-specific instances of religious piety and solidarity are being lost.

Sustainability presupposes the valorization of the culture and history of places and the involvement of local communities whilst at the same time avoiding forms of cultural or economic colonization: even the super-rich coming from cities or other continents should see themselves as guests to local communities as opposed to envisioning them as anonymous hunting and fishing reserves. It bears better fruits to focus on the development of instances of local micro-entrepreneurship rather than on excellent, yet exogenous finance.

Reactivating Cultural Processes to Engage the Territory

Synthesis by the authors of the research colloquia with Stefano Della Torre, Professor of Architectural Conservation, Politecnico di Milano, Department of Architecture, Built Environment and Construction Engineering.

Systematization and fruition, both of a physical and digital nature, of the rich and dispersed cultural offerings of marginal territories is key to their regeneration. The whole idea of integrated cultural systems had already been present before 2000, but limited to instances of intra-sectoral integration (e.g., networks of museums or archaeological sites). At the beginning of this Millennium, the concept of "Cultural District" (*Distretto Culturale*) emerged, which paved the way for the value-added of engaging with one's territory. As a consequence, integration turned out to be of an inter-sectoral nature (in-between culture, economy, and society: e.g., a museum is no longer deemed as a sealed cocoon, but rather finds its *raison d'étre* in the cultural production it expresses and the income it begets, also in terms of social involvement. The productivity of the cultural system is planned as part of a supply chain involving diverse productive sectors, not solely tourism, with the main objective being the creation of a creative ecosystem capable of improving employment opportunities and life quality (if a territory as a whole becomes more attractive, people will live there more willingly).

Success of the Cultural District initiative has been quite patchy throughout Italy. In Lombardy, the Fondazione Cariplo banking foundation allocated 20 million Euros between 2009 and 2013, to which 40 million was added as co-funding, with the latter addition testifying to the success of such initiative. A series of interventions were bankrolled, most of which entrusted to local businesses; territorial and local-authority engagement was real and tangible, thus proving the Cultural Districts model as capable of begetting sustainable projects in the most meaningful sense of the term.

The "Cultural District of Valtellina" (*Distretto Culturale della Valtellina*) initiative has been among the most fruitful experiences, capable as it was of integrating local wine-industry excellences on the one hand and cultural landscapes on the other for the purpose of de-seasonalizing tourism and re-balancing the Mid- and Upper Valtellina sections.

Even before the COVID-19 pandemic, some evidence had emerged of growth in off-season tourist attendance and the achievement of a balance between summer tourism and the already-established winter tourism. Contexts such as the Mid-Valtellina, which have been on a path to diversify their tourism offerings for about two decades, have actually been benefitting from climate change.

The Route of Terraces (*Via dei Terrazzamenti*) (2013) is the main driving project of the Cultural District in light of its capacity to engage the territory, much more so than other projects (as in the cases of the auditorium in Morbegno or the Tibetan bridge in the Val Tartano). The Route of Terraces is the backbone of value creation for all the locations it links to one another. Indeed, it triggers deep participatory processes, multilingualism, and fosters local resident awareness and empowerment.

The Cultural District model implemented in conjunction with Fondazione Cariplo in Valtellina in the years around 2010 has its main theoretical and experimental basis in the research by Stefano Della Torre's group at Politecnico di Milano. The precursor of the district model is the Halland Model (Christer Gustafsson, Uppsala University), an experiment performed in the 1990s in the homonymous Swedish region whereby anti-unemployment funds were successfully employed to fund restoration works on one hundred degraded historic buildings, thus allowing for the unemployed to be trained and empowered as opposed to supported by way of governmental subsidies; the project was carried out thanks to a cross-sectoral agreement linking the unemployment office, the Superintendence, and the construction company syndicates. Not only did the outcomes encompass the provision of jobs, but also the training and social empowerment of unemployed subjects, many of whom were migrants, as well as, obviously, the much-needed restoration of historic buildings. In more recent years, the mainstays of the Cultural District concept have been shared with other significant experiences developed worldwide, including the prototype-project of the Monte hamlet occurring within the Municipality of Castel San Pietro in Switzerland (Dieter Schurch, LISS—Laboratorio di Ingegneria dello Sviluppo Schürch), the Cuenca, Ecuador experiment (Fausto Cardoso Martínez, Universidad de Cuenca), and the Piedmont-based eco-museums (Maurizio Maggi, IRES—Istituto di Ricerche Economico Sociali del Piemonte). This original and innovative path of theoretical–experimental research rests upon cross-field skill acquisition rather than on the attaining of short-term, merely tangible results, all of which may well be summarized in the following slogans: "It is not so important to have a restored castle but a person who has learned how to restore it"; "One person cannot build a house alone, but 10 people build 100" (Hassan Fathi).

It is thus crucial to re-activate local-community cultural dynamics. So what do these local inland-communities look like today? The percentage of natives is similar to that of immigrants and/or tourists in the high season, a fact even the few parish priests still occurring within small towns are aware of, witnesses as they are to the endangered intangible cultural heritage surrounding them and custodians to many a half-abandoned church in Italy's marginal areas. The small churches and chapels characterizing the historic landscapes of the Italian countryside and mountains can only be saved if the projects aimed at their restoration and reopening act as catalysts of mutual collaboration. One of the main objectives of the reactivation of the culture of ever-evolving locals is to foster community awareness and pride among children and youth. A few flexible legal tools may prove useful to this effect, as is the case with Community Cooperatives (*Cooperative di Comunità*), a format employed to re-activate services in the mountainous village of Roncobello (the Province of Bergamo), which in turn poses a typical example of how the publicly-funded recovery of an abandoned school may serve as a catalyst for the revitalization and innovation of socio-cultural relations.

In this perspective, technological innovation may be of great help, as in the case of the experimental project "Open-door Churches" (2022), which through a free app allows to independently open and visit the ecclesiastical cultural heritage sites of Piedmont and the Aosta Valley. To beget some actual value, our cultural heritage needs to become disengaged from volunteerism, and be managed under the realm of well-designed site- and culture-specific programs instead.

It is of utter importance to intercept processes of trend reversal, to tell individual success stories of life and work in towns "to be reborn", to mention young-age witnesses of courageous choices who have found it in their hearts to live in isolation without feeling cast out, but rather embracing a "glocal" lifestyle; young people well aware of being in the process of expressing new trends (trend-setters), and capable of re-interpreting traditional activities with contemporary values and opting for them "in spite of" their high-degree study curricula, all of which proves very useful today for the rural activities of livestock farming, pastoralism, agritourism, focused on the organic approach, zero-km, and the integration of products and services in an experiential sense.

Salvation rests with cultural reactivation and proceeds, among other things, from the training of and listening to locals as well as authentic social dialogue, both of which to be pursued regardless of one's place of origin (mountaineers vs. metropolitan citizens); even those who choose to belong to a place regardless of the timespan of their physical presence as smart workers, digital nomads, or holiday-home owners, may well lend a valuable contribution to the building of communities united by a sense of belonging, albeit heterogeneous in nature; communities that express creativity and entrepreneurship at once, and a shared culture both well-rooted and vital.

The Idea of the "Route of Terraces" (*Via dei Terrazzamenti*)

Synthesis by the authors of the research colloquia with Dario Foppoli, Technical Director of Foppoli Moretta e Associati engineering company, Tirano (Region of Lombardy), and *Via dei Terrazzamenti* project coordinator.

The whole idea of the "Route of Terraces" (*Via dei Terrazzamenti*) resulted within the framework of the "Cultural District of Valtellina" (*Distretto Culturale della Valtellina*, 2009–2014) project with the aim of building a unified cycling and hiking path in a mid-altitude area running through the terraced areas of the Rhaetian mountainside and connecting historical-artistic buildings, wineries, farmhouse restaurants and lodges, and religious sites of considerable value and interest. Well-rooted in the Valtellina since the Middle Ages, viticulture and wine production can be considered an early form of proto-industrial activity in the light of both their very wide spatial articulation, organizational complexity, and economic importance, and their having been the first-ever Valtellina-based business profiting from a wide commercial network. Indeed, many a noble landowner gave their property to farmers for use, which were in turn reclaimed and cultivated to produce wines especially suitable to be carried over long distances and therefore marketed by the nobles themselves beyond the Alps and as far away as the Helvetic Confederation, Bavaria, and the Habsburgian Empire.

For this very reason, the project of the "Route of Terraces" was intended from the very beginning to form a geographic and thematic continuity with historic Swiss route No. 30, to wit, the "Via Valtellina" connecting Tirano to the Rhine Valley and not surprisingly promoted as the route along which Valtellina wine was forwarded to Central European markets by the so-called *somieri* (beast of burden conductors who carried wine in barrels). The long-term vision was to connect the "Route of Terraces" in Italy with the "Via Valtellina" in Switzerland, and Lake Como with Lake Constance, thus tracing a slow-tourism corridor of European relevance coinciding with the ancient wine route: from the terraces where the vines were (and still are) produced, all the way to the Alpine passes through which the wine was transported.

The concept of rediscovering a historic transalpine corridor, already inherent in the initial idea of "The Route of Terraces", is revived in a contemporary key by this book as the Sport&Heritage corridor, a legacy of the Milan-Cortina 2026 Olympics.

The "Route of Terraces" was proposed by a multidisciplinary working group established by the Local Development Foundation (*Fondazione di Sviluppo Locale*, an emanation of the provincial administration of Sondrio) for the purpose of developing the "Cultural District of Valtellina" project (co-funded by Fondazione Cariplo banking foundation). This group, in which Engineer Dario Foppoli was entrusted with the task of coordinating activities related to tangible cultural assets, opted for focusing on the "Route of Terraces" as the driving-force project of the local territorial system. The proposal was submitted to and approved by Fondazione Cariplo, and subsequently implemented.

Development of a prioritized strategic project was of particular use in that it allowed for community team-building at a political, administrative, and technical level: in order to get a given area to become involved, one needs to submit to stakeholders a tangible project. For the "Route of Terraces", dozens of meetings were in fact held with all the main stakeholders for each one of them to identify and express their peculiar area of interest within the overall project: from local administrators all the way to cultural and tour operators, everyone was provided with a chance to identify with the project as a whole precisely in that it was of a tangible nature. The project soon manifested its utter potential to involve and coalize the whole course of the Valley from Morbegno to Tirano and foster collaboration and synergy building among the three Mountainous Communities of Morbegno, Sondrio, and Tirano on the one hand and the Provincial Administration on the other.

The implementation stage of the project extended over 3 years—from 2010 to 2013—and encompassed participatory, preparatory, and design work as well as procurement and execution until the inauguration ceremony of the "Route of Terraces" in October 2013. All players had to proceed in a very fast efficient way, and all of them managed to be on schedule. The inauguration of the trail marked the final step of the "Cultural District of Valtellina" project as a whole. It should be emphasized in retrospect that the Foundation missed an opportunity to contribute to the implementation of such activity with additional (albeit reduced) funding, which at an early stage would in turn have been proven of strategic relevance.

A few months upon inauguration of the pathway, the "Cultural District" project was terminated, and the issue of the subsequent management and promotion thereof persisted. In other words, the districts program allowed for implementation of the hardware, but did not provide any software. Moreover, the timeframe over which the project was implemented happened to overlap with the nationwide revision of the functions and resources of the all provinces, in our case the very driving force of the project as a whole. As a result, the management of the "Route of Terraces" fell within the purview of the Mountain Communities, a group of entities characterized by nature by a rather near-sighted perspective of the territory surrounding them. Among the consequences thereof is a total lack of brand coordination, as became evident from the fact that in the context the most famous local sporting event, a marathon taking place along the "Route of Terraces", that name does not occur to begin with, with the expression "Wine Trail" being used in its place.

We should now draw some attention to the previously mentioned proto-industrial character of Valtellina's wine production: the idea of "walking counseling" that this book suggests by way of the Smart Terrace Trail project and aimed at instances of strategic management consulting during walks along the "Route of Terraces", is also consistent with the valorization of heritage and rediscovery of the historical value of this route. It is also confirmed that there are either unused or underused buildings in the area that may be employed for the creation of service hubs occurring along the "Route of Terraces", to wit, Palazzo Besta and the hamlet La Curta in Bianzone, Palazzo Andres in Chiuro, and Palazzo Odescalchi in Berbenno. Restoration works are either in progress or have been carried out as of recent on some of these buildings the enhancement of which might rest on the re-framing of their use as hubs of the Smart Terrace Trail to include residences, coworking spaces, restaurants, etc., with the added benefit of positioning them as optimal subjects in the context of upcoming calls, e.g. by Fondazione Cariplo and for the sake of continuity with the initiatives previously implemented under its aegis.

For the sake of argument, we ought to keep in mind that the most profitable activity in the Mid-Valtellina has historically been that of viticulture: however, this is a non-industrial production model sustained by people who, especially in the light of the de-population phenomena of the area and the current demographic crisis, are gradually turning older. The current relinquishment of terracing is very difficult to counter, and the only vineyards where production is still sustainable are those occurring within the perimeter of the DOC production area that ends in Tirano: just beyond that perimeter, all terraces have suffered dereliction over the last half century. Now that the importance of terracing is better understood, the human capital capable of maintaining and putting them to good use is missing.

To outline a model for revitalization for the mid-mountain (i.e., valley floor and slope territories with an altitude below 1000 m) is an utterly complex task, as the nationwide dynamics of de-population and economic crisis are most heavily perceived in these very territories. It is way easier to harbor an optimistic outlook from the perspective of a metropolitan resident than from the standpoint of the people of a valley.

However, it must be pointed out that some significant initiatives are being developed in the area. For example, work is underway in Sernio, the first municipality upstream of Tirano, to rehabilitate 4 hectares of abandoned terraces, where alternative crops such as olive and hazel (which have much lower cultivation costs than vines) or small fruits (which cost money but have shorter time frames than vines) have been introduced. On the castle hill of Grosio, vineyards have been re-planted, and winemaking re-instated in collaboration with the Fojanini Foundation (Center for the enhancement and empowerment of applied scientific research in agricultural disciplines and technical assistance in the Province of Sondrio). However, these exemplary initiatives, co-supported by Fondazione Cariplo as part of the "emblematic projects" (*progetti emblematici*) funding scheme, do not seem sufficient to generate full sustainability, thus bearing the question of how any further value added might be attained.

The increase in tourist flows, highlighted at a local level especially in the area of Tirano in the light of the inclusion of the Rhaetian railway in the UNESCO heritage, does not seem destined to leave a substantial imprint on the welfare of the territory. The experience of the Cinque Terre (also a UNESCO heritage site) teaches us that despite the current increase in tourist inflow (i.e.: overtourism) the territory in itself does not reap any benefits: even in the Region of Liguria the terracing has been falling victim to relinquishment, with the ensuing orographic disruption causing disastrous landslides.

More work on residency patterns is needed to address this situation: the problem does not relate to the availability of jobs, as current technologies allow people to work from anywhere in the world. The difficulty rather rests with the fact that moving to the mountains (even just to mid-mountain areas) on a permanent basis presupposes a mindset capable of adapting to the "spirit of the place", which is consistent with long time cycles, respect, acceptance, and perseverance in relations with the environment, whether in a natural, social, or business perspective. The attraction of temporary inhabitants to the mid-mountain, albeit capable of eliciting some positive outcome, does not beget any real impact, in that true long-time planning is paramount to really affecting the recovery of a territory and cultural landscape.

Digital Communication Technologies for Inland Areas

Synthesis by the authors of the research colloquia with Pietro Guindani, Fastweb + Vodafone.

Digital communications can help preserve the value of inland areas by giving them equal access to services, especially in Italy with its unique cultural heritage, rich in history and traditions of human, social and artistic value. Italy has 5521 small towns, accounting for 69.9% of its 7896 municipalities (IFEL on Istat data, 2024), about half of which located in inland areas, often with reduced accessibility to essential services.

New communication technologies offer a real opportunity to revive Italian villages because they allow us to overcome the challenges of geographical constraints and complex terrains by facilitating access. Connectivity in inland areas is improving significantly in terms of both capacity and territorial coverage, thanks to operators sharing their infrastructures, as well as the enhancement of new technologies such as 5G FWA (Fixed Wireless Access) and national and European public support, such as the PNRR. These developments encourage many small municipalities to plan actions to reverse the depopulation trend. In fact digital coverage is not just a solution to an infrastructural problem—it's boosting new opportunities and providing digital solutions that can enhance the cultural and natural heritage of villages by promoting tourism with innovative tools such as interactive apps, digital maps and augmented reality content, thereby driving local economic initiatives, and attracting professionals and companies looking for places where they can have a better work-life balance. In this sense difficult geographical terrains are no longer an unsurmountable problem—in fact they can become the attraction. With increasing digitization Italian villages can become the protagonists of a rebirth, combining tradition and innovation to build a sustainable and inclusive future.

Reliable and widespread connectivity makes it possible to work remotely, attracting new families, young professionals and entrepreneurs to villages where there is often a lower cost of living, a higher quality of life, closer connection with nature, and a more relaxed lifestyle, without having to give up certain job opportunities that would normally only be available in the city. At the same time digitization would lead to the redevelopment of physical spaces within villages. Old public buildings or unused spaces can be transformed into modern, functional coworking areas designed to accommodate remote workers and startups. In this way villages can attract digital professionals, contemporary artisans and small businesses, who see in these places a combination of high quality life-style and essential technological infrastructure.

Digital connectivity is also a strategic lever for innovative services, such as distance learning, telemedicine and digital management of bureaucratic practices. These tools improve the quality of life in the villages, making them not only more attractive for those seeking a change of life, but also more functional for the current residents.

In this sense digital remote working is not just a booster for more job opportunities: it becomes a catalyst for structural, social and economic change. If supported by investments in connectivity and territorial planning, it can trigger a virtuous circle that enables villages to reverse demographic decline, revitalize its economic and social fabric and contribute to a more polycentric and sustainable development model. New job opportunities can reverse the trend of depopulation, which in recent years stands at about 73% in towns with less than 5000 inhabitants (National Association of Small Municipalities (ANCI), Atlas of Small Municipalities).

Inland areas are mistakenly considered marginal in terms of income generation, but if they have the right resources and tools, they can become protagonists in competition between urban and internal territories, thus creating an ecosystem in which innovation, digital infrastructures and human capital converge. This is the only way to unlock their potential and transform them into new poles of growth and economic, social and technological sustainability. Digitalization offers smaller communities the means to participate in global economic, cultural and social networks without having to compete directly on urban density.

The digital transition, therefore, should not be seen as exclusive to cities, but as an opportunity to rebalance development between urban and inland areas. In this perspective, the future of inland areas will depend on their ability to seize the opportunities offered by new technologies, transforming historical limits into competitive advantages and building a more inclusive and sustainable development model. The inhabitants of small towns (58%) compared to residents of large cities (44%) seem more concerned about the lack of digital services, seeing this as an obstacle to sustainable development (Fondazione per la Sostenibilità Digitale, 2024). In fact, the village is the ideal place to design and build the same infrastructural conditions and services as the city, but in a more sustainable ecosystem that offers a better quality of life, and therefore a longer life expectancy, and fosters a sense of community and mutual support in daily life.

From this perspective the 5G radio-mobile service and fixed broadband represent much more than a technological evolution, because they can transform Italian communities into connected and sustainable spaces, able to face the challenges of the present and seize the opportunities of the future. These technologies are the basis of the "Gigabit Society", with a vision for fast and reliable connections to enable innovative services and improve the quality of life for citizens and administrations.

Experiential tourism for example can benefit greatly from digital strategies that make the uniqueness of these places accessible: broadband and 5G can enhance the cultural and landscape heritage of Italian villages, allowing them to enrich their offer and improve hospitality for tourists.

As well as tourism, 5G can contribute significantly to improving the quality of village life, enabling innovative services also in the health, education, mobility and security sectors.

Outside large urban areas telemedicine and remote monitoring can improve access to health services and allow direct and effective communication between doctor and patient, with high-quality video and real time digital sharing of clinical data, an improvement in access to health services for residents of inland areas, reducing timescales and offering continuous support even remotely.

Even small towns can benefit from 5G solutions for smart cities: video analysis systems make it possible to monitor traffic, prevent gatherings and detect natural phenomena such as floods or landslides, to ensure safer and more efficient management of the territory.

Digitalization also brings benefits for the security of our artistic and cultural heritage, by protecting the environment and visitors with AI and video surveillance; the Behavioral Analysis digital application prevents break-ins and vandalism, while Online Museums keeps them safe.

From cultural enhancement to safety, from health to smart mobility, with innovative services for citizens and tourists, 5G and broadband represent the key to achieving a vision of inclusive and integrated change.

As stated in the "Borghi Avvenire" dossier of Legambiente 2024, today it is necessary to "invest in urban regeneration and social innovation with new personal services and new spaces for sociality, new models of living and tailor-made digital citizenship services, new forms of residential work based on green economy and energy democracy with renewables" (Bonfanti, 2024).

Art & Community for Regeneration

Synthesis by the authors of the research colloquia with Salvatore La Spina, Mayor of the Municipality of Centuripe (Region of Sicily).

The town of Centuripe (5000 inhabitants on the Sicilian hills, in the Province of Enna) is experiencing a serious process of depopulation. The locals are highly educated, but after middle school some of the kids typically go to the local professional hotel school, while those who want to go to high school go to Enna, Paternò, or Catania. To continue their studies until graduation, they will go to Catania or Palermo, while the young graduates will remain in the area to find a job. The young people from the hotel school instead perform internships at coastal locations where they will later stay at. The young people who stay in the town will commute to their places of work. Centuripe trains very well-prepared young people, but then loses them because the area lacks any true production, manufacturing, or entrepreneurial network. There is no generational turnover (few newborns and many demises). This phenomenon affects the entire Sicilian hinterland, whereas coastal city population is growing steadily.

The Municipality of Centuripe has aimed to recover its village, with great attention to history, to at least try to counter a de-population trend that seems unstoppable. In this perspective, and in spite of their offices being heavily understaffed, the Municipality has participated in public funding calls and obtained funds with which emblematic historical buildings such as Villa Corradini, the Dogana theater, the facade of the Cathedral, panoramic viewpoints, the rural village of Carcaci (for new hospitality) were recovered; a parking area for campers was created, too. The recovery process was extended to the surrounding environment with trekking routes in the gullies and in the eucalyptus woods. A regeneration project is in place for the archaeological area which rests on funding from the National Recovery and Resilience Plan (*Piano Nazionale di Ripresa e Resilienza*—PNRR). Public funds have been used to allow young people to open businesses such as B&Bs, guest houses, restaurants, tailoring shops, ceramic workshops (Centuripe is famous for its polychrome vases); in addition, a hotel with more than 30 beds is opening, which points to a likely trend reversal. The potential for recovering the remains of the sulfur mine and the ancient terraces surrounding the town is highlighted, which could be reused for growing vines, olives, almonds, or even saffron (an ancient local crop since the time of Plinius the Elder). Unfortunately, the ability to intercept foreign or Italian private investments is lagging, held back as it is by fear of the mafia, a phenomenon luckily absent in Centuripe.

Art is seen as strategic: an old deconsecrated church has been rehabilitated for contemporary art, a couple of national-level exhibition spaces have hosted international exhibitions on avant-garde art and the European twentieth century, alternating with local exhibitions to put young artists to work. Art exhibitions draw 1000–1500 people in the summer months; daily visitors also create induced revenue.

The Municipality is trying to create a virtuous mechanism through regeneration interventions, organization of events, promotion of the territory (participation in tourism fairs such as the BIT, documentary films made possible by regional funds, and development of the great potential of local foods and wines). Local agri-food craftsmanship, however, is archaic: the Centuripe area boasts a protected geographical indication (PGI) for blood oranges with small consortium producers, small dairy and confectionery producers (with their typical dessert going by the name of "Bersagliere di Centuripe"), small family businesses producing sausages, oil, and excellent saffron thanks to the suitable soil. Yet, none of those producers comes in the form of a consortium. There is a lack of an entrepreneurial component to sell on a large scale, such as a supply chain of products from local oranges, with branding.

We need training in entrepreneurial risk culture to stimulate brave young people: selling online would be a quantum leap. So, it is a problem of both lack of capital and lack of entrepreneurial culture.

The "Annunziata" district under the town's rocky ridge, which is very spectacular, would be perfect for a diffuse hotel that could act as a catalyst for complementary activities such as small restaurants, artisan stores of typical products, and cab drivers for local routes: a foreign investor is needed who believes in Centuripe and may trigger a true revitalization process. Some relief is also sought in foreign interventions based on the history of famous tourist destinations of Italy: according to the Mayor, the success of Taormina was not made possible by locals but by the Germans, and Florence became what we know it to be today thanks to Britons, etc. Even today, in Sicily the renaissance of the village of Reale, under Mount Etna (Castiglione di Sicilia) that has become a center of wine production of excellence ("Etna Rosso" wine) is due to the investments of entrepreneurs from Veneto and England. It should not come as a surprise that those living in a depressed area are unlikely to develop enlightened visions, and that a spark of enlightenment is more likely to come from the outside world instead. Even the choice of the village as the film set of a great director helped it, thanks to the tourism flow that arose as a consequence thereof.

These opportunities are likely, but the village must be ready to seize them. The management of Centuripe's exhibitions and sightseeing tours is entrusted to cultural associations that are beginning to get young people to work in the cultural sector, as guides. The cultural operation is aimed at featuring the village as one of the most interesting ones throughout Sicily (i.e., "the balcony of Sicily"). The town must be aware of and ready to welcome foreign tourists and investors who believe in the beauty and the small, yet fascinating treasures of Italian hamlets.

In other words, public funds support regeneration interventions, but any instance of post-regeneration management would require investment and entrepreneurial culture that can only come from the outside, particularly so from abroad. We need to start working with young people, who breathe classical culture here and the awareness and pride of being born in a special place, but lack the entrepreneurial culture and education needed to commit oneself to entrepreneurship as opposed to pursuing a government job or just leaving.

The personal story of the mayor of Centuripe is a revealing one: born in Centuripe, after going to university and a long period of work in Florence, at 49 he decided to return to his native town, out of attachment and a tad of home-sickness. Such an attitude is quite widespread among the inhabitants of the towns of Sicily who remain proudly in love with their land blessed by the gods, even if they had to abandon it to fulfill themselves professionally and make a stable living. The mayor has decided to bet on the potential of his village, thanks also to experiences outside that make possible the assessment of such potential.

The Challenge of Rebooting Mid-Altitude Mountains

Synthesis by the authors of the research colloquia with Roberto Laffi, General Director for Territory and Green Systems, Regione Lombardia.

The 2021–2027 strategy for the development of "inland" areas put together by the Region of Lombardy (a total of 488 Municipalities, located mainly in the pre-Alpine and Alpine belt, with a population of about 1.7 million), is aimed at revitalizing these areas by intervening on the supply of essential services (e.g., education, health, mobility) and promoting socio-economic development based on the valorization of local resources (of an environmental, socio-cultural, and economic nature). The planned actions also aim to encourage greater digitalization and innovation of small and medium-sized enterprises, energy efficiency, integrated tourism, and cultural services, with the addition of sustainable mobility. The topic of inland areas is also addressed in the Regional Territorial Plan, a strategic tool of regional planning currently in the process of being revised in the light of a broader context and the relations in place between neighboring territories.

The public funds allocated to support the strategy (European, national, and regional ones) are in excess of EUR 190 million. To those, we ought to add the investments earmarked for the Valtellina in the light of the 2026 Winter Olympics and the resources allocated by the Region for urban regeneration (interventions on buildings and public spaces), which in the Province of Sondrio have involved 41 Municipalities for a total amount exceeding EUR 26 million. In addition, many a local authority pertaining to any of these areas can also count on investments related to the implementation of the National Recovery and Resilience Plan (*Piano Nazionale di Ripresa e Resilienza*—PNRR).

Beside resource availability, another important issue for the revitalization of the mid-altitude mountains remains that of the development model for these areas and the integration of policies affecting them. This is a crucial step in overcoming the stigma of disadvantage and marginality in the light of which these territories, in turn not directly reaping the benefits associated with mass high-altitude winter-sports tourism have always been viewed. Of special relevance will therefore be the coordination of policies within a unified framework and the adoption of measures capable of valorizing each and every one of their vocations whether they be of an environmental, agricultural, gastronomic, or business nature.

Another truly important development opportunity is linked to the significant growth of international tourism, fostered as it is by the environmental, landscape, and food-and-wine excellences of our Alps and Pre-Alps in addition to traditional, renowned art and culture hubs. A phenomenon not new to Valtellina—let us just think of many a renowned ski resort—that is now, indeed, experiencing some further consolidation by virtue of a trend favoring travel experiences in less-known places, with itineraries linked to outdoor sports, contact with nature, and wellness (a special mention be made, for instance, of the tourist influx from the Far-East and the significant growth in the share of tourists originating from China).

From this point of view, it is worth mentioning that Valtellina can boast a significant extent of protected areas of great environmental value, including the Stelvio Park, the largest one in the Alps, the Orobie Valtellinesi regional park, and no less than seven nature reserves.

Among other lines of strength for "sustainable" tourism are the famous "Bernina Express" train, which in Tirano allows for interchange with the railroad line linking

Sondrio to Milan, the "Valtellina Trail" (*Sentiero Valtellina*), a bicycle greenway grazing the Adda River and leading from Tirano all the way to Valposchiavo, again providing interchange with the railroad, and the "Route of Terraces" (*Via dei Terrazzamenti*) on a mid-altitude area of the Rhaetian slope in-between Morbegno and Tirano, a pedestrian/cycle route extending among vineyards and wineries. All of these infrastructures have already born a positive impact on the local economy and increased the attractiveness of the territories at issue.

Finally, we cannot refrain from mentioning the great opportunity provided by the Milan-Cortina 2026 Winter Olympics and Paralympics to make the beauty of these territories known worldwide and promote, also through sports, a sustainable development not only from an environmental standpoint, but also from a social and economic one.

Territorial System of Interventions Shared by Communities & Researchers

Synthesis by the authors of the research colloquia with Pietro Petraroia, CEO of Cultura Valore S.r.l., Milan; Former Superintendent for the Artistic and Historical Heritage for Western Lombardy and the Pinacoteca di Brera (1991–1997).

In territorial regenerative processes, and especially in inland areas, it is crucial that languages (in all of their meanings) used in the relationships between researchers and residents enable projects to be co-authored, so that the population envisions them as their own and therefore wants to implement them to begin with. Innovative exogenous visions can often be a positive trigger from the outside; however, for them to last effectively, it is necessary to create a common ground of knowledge and languages, where interaction is fertile, based as it is on valid information that is truly shared, as well as on well-understood needs.

Instead, people living in the most isolated places often become fascinated by solutions that actually dissipate territorial resources, as they lack the tools and data to figure and pursue solutions more suitable for the development of their territorial capital.

An excellent reference for the methodology of building some common work between researchers and residents is the intense contribution, indeed the very life of Hugues de Varine (de Varine, 2005), which he spent in a decade-long experience of recognizing and valorizing the cultural heritage of inland areas, including intangible heritage, with a focus on the role of cultural and linguistic (in short: relational) interaction between the population with its historical—and perhaps static—experience and the researcher's fresh eye.

Indicative of this issue, among many others, is the case of the Municipality of Gerace in Calabria (population: 2400), suffering, like so many others, from depopulation and youth abandonment; to Gerace, the National Recovery and Resilience Plan (Piano Nazionale di Ripresa e Resilienza—PNRR) allocated twenty million euros[1] by decision of the Region of Calabria, to be distributed through calls for proposals that have not, however, seemed to reap any enthusiastic adherence. It cannot be ruled out that among the reasons thereto is a lack of local confidence in the ability to sustain administrative procedures, particularly reporting procedures, which may prove too complex and of uncertain outcome. The national events of the Superbonus, also known as the 110% scheme, may have influenced in too many cases a local perception that seems to have remained distrustful. In this specific case, there may be orienting difficulties, that is, picturing oneself with respect to the objective of the call for proposals,[2] which calls for the creation of an *albergo diffuso* (i.e. scattered hotel), for luxury tourists in the historic village. It says that "citizens will be called upon to collaborate and to cede with a 20-year gratuitous loan some properties to be used as a diffuse hotel". Pietro Petraroia wonders whether the formulation of the call for proposals and related projects was preceded by a phase of listening and analysis of the problems most felt by the population (with special care, also, to those unexpressed), as well as an adequate information campaign on the supports available to those who would join the call for proposals in order to facilitate, among small businesses, the smooth and successful management of the complex funding procedures. He gets the feeling that the prospect of a single managing entity to be activated (as the *albergo diffuso* would be, for obvious organizational needs) and the emphasis on the "luxury-style" orientation of the intervention may in several cases generate a certain reluctance.

In fact, much of what is envisaged by way of projects of this kind is suited for conversion into the improvement of the daily living conditions of residents; improved widespread habitability, as well as health and public transportation services, maintenance, and infrastructures, even of a digital nature, among public spaces (to wit, objectives that have somehow leaked out in the news about the "Porta del Sole" project in Gerace) could be a basic factor in avoiding abandonment by residents, generating, as a consequence, the prerequisites also for more attractive tourist reception services. An area, this one, of economic activity of interest, but too often hastily identified as the panacea, in terms aloft from territorial system interventions and perhaps not very adherent to the daily life needs of the residents, whose eviction should in fact be avoided, given that depopulation is already triggered and is, indeed, precisely what one wants to prevent from continuing.

Among the areas of intervention to promote wellbeing in a village (and thus permanence, both by residents and guests), apart from basic services, a very current strategy and theoretically well present in the PNRR programs can be identified, namely the combination of ecological transition, digital transition, and anti-seismic but also energy improvement of historic buildings, moreover to be protected as a diffuse cultural heritage, even if no individually listed buildings occur; it goes

[1] In reference to the case of Gerace, see the newspaper article: https://www.ilsole24ore.com/art/gerace-comunita-creativa-20-milioni-euro-delibera-AEiqLwPB?refresh_ce=1#U401149262538e5B.

[2] See the public call for tender: https://www.fincalabra.it/web/index.php?option=com_content&view=article&id=11564&catid=66.

without saying that in places like Calabria it can be precisely the experience of warm welcome by residents which poses the main attraction to those coming to visit.

While it has long been accepted that earthquake-proof structural improvement of historic buildings should be pursued as part of a conservative intervention (see also Article 29 of Legislative Decree No. 42 of January 22, 2004), the same is not yet true for energy improvement, for which we still teeter between the opposites of doing nothing (in the case, however, of only listed buildings, since they are currently exempted) or pursuing (for all others) an adaptation that would require such invasive interventions as to be incompatible with the requirements of the protection of the historic settlement fabric, as in the case of hamlets.

Pietro Petraroia believes that conspicuous financial support, such as the one mentioned above in the case of Gerace, could be more socially and economically sustainable in the long run, if only it generates a meta-project for the local government and citizens capable of guaranteeing a significant energy improvement that encourages both ordinary residency and tourist hospitality in both the hot and cold months. It could be based on the reactivation of the capacities inherent in local building types (both in terms of limiting heat dissipation in and out, and in terms of co-generation) in combination with more innovative technical solutions that are perfectly compatible with the requirements of protection.

Several tangible examples were presented in the afternoon session of the October 4, 2023 conference in Brescia, Italy, a symposium sponsored by the local Chamber of Commerce.[3] Without going into the details of the many proposals already available on the Italian market today, Pietro Petraroia would like to point out that several centers of ancient formation in inland areas could (or could have?) been places of experimentation and development of available and combinable solutions, which would deserve a broader market, thus able, also, to generate the necessary resources to refine and engineer many further solutions now available in research centers, but at a prototype stage.

Cultura Valore S.r.l. (which conceived the aforementioned Brescia, Italy conference), the General Directorate of Archaeology, Fine Arts, and Landscape of the Italian Ministry of Culture, the Politecnico di Milano, the Politecnico di Torino, the University of Brescia, and the Chamber of Commerce, Industry, Crafts, and Agriculture of Brescia entered into a framework agreement at the end of 2024 to focus on energy upgrading methods that are compatible with the protection of historic buildings, not so much so to provide a repertoire, an abacus of ready-made solutions, but rather to experimentally develop a procedural protocol to help deal effectively at the design stage with the most diverse cases: an approach intended to be both site and culture-specific. In fact, just as in the medical sector it is well-advised to "listen" to the individual patient regardless of there being general treatment protocols for one's disease, in this context it is appropriate to start out by turning an attentive ear to communities and somehow scanning the built

[3]The video of the Conference "Sostenibilità energetica e tutela del patrimonio culturale", held in Brescia on October, 4, 2023 is available at the following link: https://www.youtube.com/watch?v=YAZU8a-_N2Y.

environment, as opposed to imposing instances of planning and objectives from above through inflexible and aprioristic performance rules.

In fact, it would be preferable for technical solutions that can be identified through the use of "system" study and design protocols, such as those hypothesized, to be directly tried out by the individual owners of a broad percentage of housing units (people not to be "evicted"!), thus in addition avoiding tenders for large contracts, which administrations of small municipalities have a hard time managing in an expedited fashion. Moreover, such an approach would enable for better engagement of the resident population—which should remain the direct beneficiary of the interventions, regardless—by motivating it with verifiable mid- to long-term bill saving perspectives and supporting it with the necessary design supports, administrative and fiscal advice, as well as through well-devised authorization systems.

It should be borne in mind that experimenting in small, inland-area municipalities with solutions of this kind may allow for the refinement of procedures and methods suitable for being transferred to portions of cities that are not depopulating, while nonetheless valorizing the entire construction industry and related handicrafts.

In short, the innovative entrepreneurship of a single private individual who joins a community project should be stimulated and supported, without leaving him/her alone, but rather by flanking him/her with public support, including consultancy and technical-administrative assistance: for example, a highly persuasive guarantee for the investment of a single private individual on his/her own historic building would reside in a framework agreement priorly authorized by the local Superintendence of Archaeology, Fine Arts, and Landscape, and, first and foremost, in the one's being offered professional assistance for the harmonization of the project with the actual bureaucratic procedure.[4] This is just how the P4 may come true in its shape of a private-public-people partnership.

Walk the Change

Synthesis by the authors of the research colloquia with Serenella Sala, Founder of Serenella Sala & Associati—Executive Counseling Services, Milan.

Walking has the power to become a tool of personal evolution: it increases our introspective ability, sets us in a context with others and with the surrounding environment in a different way than we are normally used to. While new perspectives, certainties, and behaviors open up new paths within ourselves, walking reveals itself as an active metaphor of this transformation. Serenella Sala, in the framework of her executive counseling services in Milan, put together a program that combines walking with more traditional ways of personal and professional transformation,

---

[4]The following links are related to the ongoing discussion on cultural heritage and inland areas: https://www.agenziacult.it/aperto/politiche-per-la-cultura-per-una-migliore-normalit-occorre-realizzare-il-pnrr-con-le-comunit-territoriali/; https://www.agenziacult.it/letture-lente/sviluppo-sostenibile/la-compatibilita-fra-tutela-e-miglioramento-energetico-del-patrimonio-culturale/.

which goes by the name of "Walk the Change". It is indeed a practice that has nothing to do with trekking, nor so with the achievement of results in a sporty or competitive context. Passionate of historical walkways, she had the intuition to find an unexplored potential for professional consulting strategies inside the walking experience, where freedom, balance, and well-being live together.

Indeed, all of her clients—executives and managers all of whom burdened with responsibilities, trying to make sense of their being part of this world and looking for new ways of living—would have found the time and space of walking highly beneficial. She decided to start practicing Walking Counseling just before the Pandemic, and the lockdown pushed the idea: she reaped lots of interest, which later turned into increasingly positive comments from an increasing number of satisfied clients. Walking might indeed play a strategic role in a corporate environment: executives and managers often experience an unnatural separation between the overdeveloped and exploited cognitive world on the one hand and their emotional, relational, and bodily lives on the other. In the corporate world, we rarely allow ourselves to manifest areas of the self that, instead, are more fully recognized within private life, i.e. feeling emotions, listening with all our senses, and empathizing with others. With "Walk the Change", a pleasant and easily accessible activity such as walking and fully integrated in the company as a leader-ship development tool, the cognitive and the physical spheres become mutually connected through an unusual and pioneering collaboration.

In the case of walking counseling services it is crucial to select inspiring paths, usually across scenic and historic landscapes that are marginal enough to retain their calm authenticity, such as the Via Francigena along the Apennines or the "Route of Terraces" (*Via dei Terrazzamenti*) in the Alps. This kind of deep and in-tense experience generates memories which bind people to these places forever. The potential for rediscovering these landscapes in contemporary and sustainable ways is quite evident, and goes far beyond regional marketing as we know it. Walking counseling could become a new form of enterprise for marginal areas as demonstrated by Serenella Sala's decision to buy a rural heritage building along the Via Francigena to be transformed into the base camp for her "Walk the Change" program.

Give Young People the Opportunity to Choose the Place Where they Want to Fulfill Themselves

Synthesis by the authors of the research colloquia with Marco Scaramellini, Mayor of the Municipality of Sondrio (Region of Lombardy); Francesca Canovi, Deputy Mayor of the Municipality of Sondrio, and Alderman for Production and Events of the Municipality of Sondrio; Michele Diasio, Alderman for Tourism, 2026 Olympics, and Sports of the Municipality of Sondrio; Carlo Mazza, Alderman for Urban Planning, Environment, and Mobility of the Municipality of Sondrio.

Sondrio (population: 21,000) is the only significant dimensionally city of the Valtellina, located as it is at the center of the Lower-Mid-Valley where the bulk of the entire population of the Province of Sondrio is located (135,000 inhabitants out of a total of 180,000 throughout the Province). Moreover, all main services of a cross-municipal degree are concentrated in Sondrio: the main police precinct, the government offices, and the schools; all other public garrisons such as the Public Prosecutor's Office are reduced and even at risk of shutting down. In fact, a process of not solely depopulation, but also reduction of services is well underway. The Social Planning Office of Sondrio (*Ufficio di Piano*) manages social services for 22 municipalities. In the Province of Sondrio we can find an area-wide social services management system. This allows for optimization of resources for both personnel and money flows (including participation in public funding calls) on the one hand, and solidarity among municipalities, on the other. Social services are mainly busy with lone (unaccompanied) minors, the elderly, the disabled, and even migrants (foreigners account for less than 10% of the population and are well cared for). Services are shared, as is the case with Hospital of Sondrio. In today's society, investments in social services are increasingly relevant, especially so in mountainous areas. The Municipality of Sondrio thus plays a leadership role in the territory of the Province and pursues a strategy of collaboration especially within the Lower and Mid-Valley, in turn accounting for a conurbation of nearly 60 km.

The most significant interventions undertaken by the Municipality of Sondrio in the way of revitalizing the area in the face of the above well-established depopulation trends include the following:

- Creation of a reception and co-housing center for the intermediate SES-segment of the population in the former Enel headquarters (thanks to a ministerial call), integrated with a center for the treatment of autism. Notably, a decision has been made to focus on social residences for mid-SES individuals, as the lower tiers of the population are already cared for by Sondrio-Lecco-Bergamo's ALER offices. The Municipality of Sondrio aims to provide housing to medical doctors, nurses, law enforcement operatives, and young people overall, to wit, individuals who have an income but need some housing assistance nonetheless as the access to rental housing in Sondrio is quite difficult; despite the availability of homes, the current legislative framework does not protect landlords who will therefore refrain from letting their property to begin with: instead of pursuing some rental-related income, they will either sell their property or keep it for themselves and their families.
- Monte Salute: accommodation and training at Sondrio's former Moncucco psychiatric hospital located above Castel Masegra on a beautiful 40,000 sq m terrace overlooking the vineyards. Funding from the Region (i.e.: urban regeneration funds) will allow for basing various types of services there: a daycare center and residency for disabled individuals, some middle-class co-housing, and an ethical hostel granting an employment to young, problematic people and aimed at both tourists and regular city users. Monte Salute operates in synergy with the adjacent CPF Vocational Training Center (*Centro Formazione Professionale*) of the

Province where an e-learning hub for Politecnico di Milano's online university curricula will open starting from A.Y. 2025–2026. The Hotel School already hosted by the local CPF is also being supported in collaboration with some famous local chefs.

Young people go to study away and this can be good for their education at a national and international level; but then they will struggle to set foot back in the area because of a long-time lack of opportunities, and this poses a major problem. We are starting to see some returns for the following reasons: high life quality, low living costs, and the upgrading of fiber connectivity. It is crucial for young people to choose where to realize themselves not to be forced to leave to realize themselves (one thing is to study, while another is to choose where to live and work).

The population of the Province of Sondrio is aging (except for Livigno, with its own geography and dynamics); but since the COVID-19 period, positive signs have been noticed:

- The cabling that arrived just before the COVID-19 Pandemic allows for optic fiber connection, which represents a *conditio sine qua non* to prevent depopulation; although upstream connection of villages due to complex orography still is often not there even for cell phones, the issue will likely be overcome soon with the national satellite coverage scheme being planned;
- the consequent possibility of remote working;
- the lockdown has led to reflection on the pattern of living and thus made people newly appreciate the quality of life and outdoor leisure offered by the Valtellina.

Young returnees have a shared narrative: to manage one's daily family life is truly convenient in the Valtellina, as most essential services are within reach in a radius of 1 km; problem solving is easier because everyone knows everyone else, which in turn allows for more direct and less bureaucratized relationships; life quality. Sondrio has always been at the top of the Sole 24 ore ranking of Italian provinces boasting the highest life quality, as in the case of Belluno, until 2019 when Bari and Milan appeared at the top of the ranking, essentially so because the statistics came to include 70 additional indicators on startups and distances from universities and airports, i.e. on opportunities rather than life quality in itself. For example, Sondrio is an extremely secure city both day and night, boasts well-kept services, is tidy and environmentally friendly, and is run by a highly committed government. In essence: lots of remarkable positives provide a perspective to young people. There are young people of Valtellinan origin who have returned from other areas of Italy, but also families who are back from Argentina.

The emergence of positives for neo-tourists is also emphasized: the city is also becoming a weekend tourist destination as a barycentric base to reach Lake Como and the Bernina Train, with the possibility of enjoying vineyards and on-site cycling in the Mid-Valtellina.

These trends were most clearly seen during the lockdown (many people moved to Sondrio at that time) and were confirmed after the COVID-19.

Sondrio in the past has lived mainly on services (mention is made in particular of the nationwide lending institutions Credito Valtellinese and Banca Popolare di Sondrio), which are now reorienting themselves toward tourism, trying to meet the growing demand for accommodations thanks to the availability of hamlets and rural buildings with potential for reuse for the purpose of hospitality and catering.

The real challenge is therefore to give young people a new perspective for them to have freedom of choice as to where to realize themselves: to provide the conditions for them to choose Sondrio for a better future. The problem is, also, of a cultural nature: a person needs to enhance one's actual potential but this potential is more apparent from the outside than from within; traditionally, there is a lack of subjective confidence in one's own territorial potential, a somewhat resigned attitude that is fortunately in the process of being overcome.

The Municipality of Sondrio is committed to intercepting the funding opportunities offered by public calls with project proposals. In order to spend the funds, the main problem is related to the public budget scheme imposed by the national legislative framework implying that ordinary maintenance be considered as part of current spending and not as investments, as would be logical, both at an urban and territorial level. The Valtellina Law allowed for large investments in infrastructure, especially in the hydrogeological sector, which proved extremely fruitful; but now the law, with its exceptional funding, is closed and the maintenance of all this infrastructure enjoys no financial coverage whatsoever. One can only rely on funds from hydroelectric concessions disbursed to Municipalities and Mountain Communities by the Region through the Province.

There is a program to improve social infrastructure that includes the construction of an ice skating and climbing arena in Sondrio which can also serve to host conferences (1700 seats + 3500 seats in the summer); 30 million euros are needed to build it and the State and Region are well on their way also in the wake of the Olympic gold (2018) of sportswoman Arianna Fontana, who by the way is from Sondrio. It is a service of excellence offered to young people to practice mountain sports at a professional level in the center of the Alps (following the example of Chiavenna where the roofing of the ice rink has borne some fruit). Covered facilities are very useful in the mountains to ensure continuity in sports practice for young people.

Regarding the effects of depopulation on the historic landscape, we distinguish historic cores that are still inhabited such as Ponchiera, Sant'Anna, Triangia on the one hand, and hamlets such as Colda on the other that are, however, becoming devoid of stores, cafes, etc. all the way to totally abandoned hamlets such as Contrada Gualtieri in the Municipality of Sondrio or Contrada Scilironi in the Municipality of Spriana, in addition to others that are lost in the woods above Sondrio, and the "Route of Terraces" (*Via dei Terrazzamenti*). This complete abandonment of hamlets with great heritage value also depends on the excessive fractioning of real estate ownership that would require *ad hoc* legal instruments. In addition to the challenge of finding sustainable reconversion hypotheses. The untended forest advances into the meadows and encompasses the ancient abandoned terraces above the elevation of the mid-altitude villages on the Rhaetian slope. There are some terraces also on the Orobic slope to be reached only by mule tracks, historic paved paths that accommodate the runoff; their abandonment causes their progressive destruction with a subsequent loss of hydrogeological safety as well as historical and cultural heritage; an intervention in such instances could be performed only on an individual basis due to its complexity. From a national policy perspective, there would need to be a decisive orientation toward the problems of hamlets where even overlapping constraints limit private intervention. The Class A energy performance requirement, which dominates the housing market poses an obstacle and represents additional costs in the sector of historic building rehabilitation. The combination of constraints and energy requirements makes the widespread rehabilitation of the historic building stock a daunting task.

The Olympics are an opportunity and not a problem, especially for the Mid-Valtellina, which does not have Olympic events to host at the risk of tensions with the local community, as happened in Cortina. For the Mid-Valtellina, the Games are an opportunity for visibility, high-level networking, and the solving of infrastructural nodes; above all, they offer a medium to long-term legacy perspective aimed at supporting segments of the economy that re-attract young people, promote research on slope exposure at these altitudes and latitudes to experiment with new rural production chains (such as olive or high-altitude viticulture), and foster novel instances of professionalism.

The Olympics add a showcase at a time that is nonetheless very opportune for areas like the Mid-Valtellina because of the shift in tourism that is taking place globally: the trend leading beyond traditional ski-focused mountain tourism. To wit, today's tourists are looking for cycling, trailing, hiking, and running. The introduction of the electric bike in the user-friendly model about 5 years ago that has opened up many possibilities, particularly for riding forest roads that allow one to go from an altitude of 200 m to an altitude of 2500 m. Another very important change has been that of short vacations and long weekends and the emergence of communities of sportsmen and sportswomen who want to explore the area in conjunction with food and wine tasting experiences, a sector which is developing fast. The Olympics could also provide an opportunity to overcome Valtellina's traditional cultural parochialism, for example by promoting the golf course and the small airport at Caiolo as amenities for the entire Valley. Sondrio is well-committed to systemizing the area's resources. On the subject of wine, it is worth mentioning the two-year project going by the name of "Sondrio city of wine" and based on a grant awarded by the Region in 2022 through the call for Trade Districts (*Distretti del Commercio*), where the Municipality is the lead partner in collaboration with others, including the Wine Consortium (producers and wine-growers from all over the Valtellina). A rediscovery of historic wineries (mostly occurring along Sondrio's Via dei Palazzi and Via Scarpatetti) is underway. The Valtellina area has been recognized by "L'Espresso 2025" renowned tour guide as the best wine-growing area nationwide. The Municipality of Sondrio supports organic businesses by promoting territorial production and marketing chains with the Organic Fair (*Fiera del Biologico*) (offering integrated experiences with local products for experiential tourism). The new Valtellina Organic District (recognized by the Ministry) has been assigned a 30% growth perspective over the next 5 years (2025–2030). The Wine Trail along the

"Route of Terraces" (*Via dei Terrazzamenti*), with passages, also, through wine cellars, had as many as 4000 competitors from 52 nations in 2024. Also noteworthy is the success of the Festival of Parks (protected natural areas) instated in 1987 and typically held in Sondrio every mid-November. Events are of critical importance, as they provide notoriety to the Municipality and the entire Valley. The strategic role of the capital Municipality is precisely that of sharing a new vision for Valtellina that frees up opportunities for young people.

New Forms of Economy for Mountains

Synthesis by the authors of the research colloquia with Gianni Verga, past Alderman for Territorial Development, Regione Lombardia and City of Milan.

A change has been emerging in the realm of mountain tourism. Until a decade ago, fashionable ski destinations used to be favored. Gradually, though, there have been signs of a shift in interest toward an alternative approach, and today the phenomenon is quite interestingly evolving in the way of sustainable territory regeneration.

There has been a resurgence in mountain agriculture (such as viticulture, even so at altitudes of 1000–1200 m) enriched by a comeback of traditional specialized organic productions (typical cheeses such as the Bitto and cereals like buckwheat in the Valtellina), and by an experimentation with produce novel to the area (e.g., saffron, olive, hemp) as a consequence, also, of climate change. Mountains are the subject of increasing interest during intermediate seasons mainly thanks to cycle-tourism: mountain biking was introduced as early as the 1990s, but the real breakthrough came a few years ago with the spreading of electric cycling within mountainous regions. Dedicated facilities such as bike-hotels sprang up in response to a demand initially coming from central and northern Europe to experience the locally pertinent legs of the "Giro d'Italia" circuit, with all of which being rounded up by dedicated bike shipping and pickup services at train stations. Later on, specialized rental services, too were made available to rent bikes on-site. For the past few years, a cycle-hiking trend has been developing, which has been attracting tourists from all over the world willing to explore newer and newer routes even in smaller, hitherto forgotten side valleys.

Nowadays, there is a widespread search for authentic, uncongested areas and places. This cultural trend also influences locals and fosters reuse of their high-altitude huts as complementary dwellings for family parties and short stays—especially so, yet not solely in the summertime (e.g., in Valtellina, see the case of the Val Grosina or the Val di Rezzalo; in South Tirol, the recovery of many "tabià", i.e. wooden barns once used for livestock and hay, near Andraz and Fedare-Giau; in Piedmont, the revival of mountain pastures upward of Rima, in Valsesia, on part of locals and citizens passionate about hunting and fishing); some cases of reopening of mountain pastures for summer grazing can also be observed (as in the case of the

Alpe Pedroria and Alpe Madrera, an initiative supported by the National Trust for Italy—FAI).

The COVID-19 pandemic has changed our habits dramatically by making us become accustomed to remote-working, which in turn has made possible longer weekends throughout the year. Rather than pursuing permanent relocations, we thus ought to focus on this very scenario: short stays and/or de-seasonalized temporary housing. Interestingly, the pandemic has borne a much greater influence on the rediscovery of a new way of experiencing mountains than the annexed co-factors of technological innovation and digital transition.

Of note is Terna's[5] program to lay high-voltage powerlines underground throughout the whole Valtellina, which would beget landscape regeneration effects in line with the novel cultural sensitivity pervading historic landscapes.

Also important is the spread of local volunteerism to organize excursions and hikes aimed at the discovery of heritage, glaciers and their evolution, and the fauna and flora. The quite recent professional figure of "mountain guide" (with an entrance exam and recognized register) plays a strategic role in this perspective. Cultural mediation leads to human re-appropriation of once anthropized, yet later abandoned landscapes, supported by a new generation of apps, tourism publications, TV documentaries, and communities of mountain enthusiasts focused on sport and/or cultural themes.

The Milan-Cortina 2026 Olympics are going to consider ski mountaineering as an Olympic discipline for the first time ever, which supports the vision of sustainable tourism in more intact and less infrastructure-rich mountainous areas.

The ski mountaineering competitions will be held in Bormio, Valtellina; also in Valtellina, Livigno with its Olympic swimming pool offers a high-altitude (2000 m) sports training hub useful for various disciplines; in light of the current climate change trend, plans have been made, also, to keep the 3000 m glacier slopes open for skiing not only in summer, but also in winter.

As for technological innovations, they should be fostered by the public sector until the market supports them. For mountains, the issue of robotics is very important for forest and water cleanup and agricultural works, especially so on steep slopes, to prevent hydrogeological risks and workplace injuries, as is the role of satellites and drones to enable high-technology supported agriculture (from harvesting-timing assessment to fertilization, all the way to operations within orographically complex areas).

With regard to mobility, climate change suggests opening the Alpine passes also in winter, particularly the Stelvio Pass connecting the Valtellina with the Val Venosta, which would pose some notable potential for alliances and connections between skiable domains of conterminous valleys.

The issue of energy is interesting due to its underused resources: in particular in Valtellina there is considerable potential for geothermal energy (Bormio, Val Masino) that could be used not only at spas but also for district heating (buildings,

[5] TERNA S.p.A. is the national company managing the power infrastructure grid.

neighborhoods, and entire hamlets). This solution could be strategic especially for Bormio and other villages where no methane gas grid is in place and heating systems still run on oil.

Also of fundamental importance is on-site educational training, especially so in the case of middle managers; it should be re-oriented to the new forms of mountain economies discussed above: so far, high schools specializing in sports have been especially aimed at the ski-tourism economy, but we would need technical institutes aimed at spread of the technologies mentioned afore.

Flower Revolution: The Alternative Farmhouse Tourism

Synthesis by the authors of the research colloquia with Giacomo Zangrando, Farmhouse Damos, Valle di Cadore (Dolomites).

The Damos farmhouse is located in Cadore on the border between the Municipalities of Valle and Pieve, about 750 m above sea level, in a sunny, yet isolated position. It is the village of origin of the family that draws its surname from the name of the native village and has kept its property over time; for some time, they made a living raising goats and cows, and the products, especially so in the case of milk, were carried on foot with barbells to the nearest village, which is 7 km away. The spatial context of the Piave River valley has been in a condition of de-population since the postwar period. In 1963, the Vajont dam disaster further precipitated the situation; as a consequence, the population moved to the valley to look for work near the eyeglass factories, a traditional sector of the Belluno area but concentrated in Longarone following the funding aimed at the reconstruction of this town destroyed by the wave of water and mud caused by the landslide that affected the Vajont basin. The hamlet of Damos was abandoned in 1966 as a permanent home and kept only for the purpose of summer vacations. In 2015, the hamlet was heavily damaged due to a fire, the causes of which were never clarified. That is when a decision was made as to whether to abandon it altogether or reinvest in it: at the age of 28 Giacomo, heir to the family, decided to take over and rebuild the property; with the support of his parents, he established the farm in 2017. From May 2024, the farm became a farmhouse-tourism center appointed with three double-guest rooms. The farmhouse-tourism center is now inhabited by the owner's family and open all year-round. In addition to the above main building, the hamlet includes a barn, a small, Fourteenth-Century church boasting some historical-artistic value, and an adjoining cemetery: the small church is still consecrated, always open although celebrations in it are rare. The barn was a logistical base during the Partisan Resistance; today, the Damos farm intends to reuse it and assign it to the recently authorized Educational Farm.

The originality of the farmhouse-tourism center resides in the choice to focus on flower cultivation, following the example of some farmsteads located in the Region of South Tirol where, however, it represents an activity adjunctive to animal husbandry and breeding. Here, instead, it is the main activity. They mostly cultivate edible flowers, and do so on the basis of the specific professional training they received at the renowned Sanremo School. They also have a drying workshop. The flowers characterize the menu offered for catering purposes, and are sold either fresh or dried to local restaurants; in addition, they have created their own cosmetic line in cooperation with local pharmaceutical laboratories.

The on-site production cycle includes organic seed selection (buying them from certified organic producers), greenhouse planting, field transplanting (partial electric fence for deer defense), summer food court with battered flowers, as well as fruits and vegetables from the farmhouse.

The agritourism business benefits from synergy with sporting-cultural activities: the agritourism is located on the "Cammino delle Dolomiti" that winds around the Belluno Pre-Alps, particularly on the stretch of historic trail between Valle and Perarolo on a promontory overlooking the confluence of the Boite and Piave rivers (a junction of the ancient flow of timber at the time of the Republic of Venice, which in turn houses a dedicated museum in Perarolo); paths used by partisans who had blazed the ancient Cavallera road there also intersect there. The farmhouse-tourism center organizes events such as presentations of books on local history and films shot on-site, lunches for locals and tourists on the occasion of the ritual of transhumance, and has equipped a climbing crag on their property which is open to guests and also used for training by the Italian IRS Police (Guardia di Finanza) as well as mountain rescue crews (great care is taken in the relationship with local rescue groups). There is also a free parking area for campers in a meadow occurring within the farm.

On the public policy side, there is some support from regional grants (for opening of new farms by young people under 30 by way of renovation loans); yet there are also some bureaucratic hurdles: the indoor catering started with the farmhouse in 2020 as a home-restaurant; after the rooms were inaugurated in 2024, the restaurant is no longer accessible due to insufficient ceiling height of the indoor room; as a consequence, the owners ended up employing a food truck to respond to customer demands.

The ideal target audience consists of middle-aged foreign tourists either in couples or families, with pets being welcome. And so are tourists looking for summer and winter-time excursions by bike, on foot, or snowshoes, to enjoy the landscapes and culture, as are also digital nomads (the rooms are equipped with large desks, and the multifunctional common room serves as both as a coworking space in addition to being a cosmetic showroom). The Wi-Fi signal strength and speed are excellent. Besides being less than an hour's drive from Cortina d'Ampezzo, Damos is also a destination in itself because people are craving authentic and quiet places more and more. The farmhouse-tourism center is also popular with locals from the entire valley.

Open as it is all year-round, the center sources most of its revenue from rooms (including some breakfast and a picnic basket). They use the AirB&B platform for ease and convenience of management. In 2022, they obtained Legambiente's highly coveted Green-Flag recognition. They have enjoyed media support even at a national level (e.g., Italy's official RAI TV broadcast), but rely mostly on word of mouth and the direct experience of a few journalists, and do not currently have an advertisement budget.

The business is family-owned with the following division of roles: young Giacomo grows and prepares the battered flowers, his mother Lidia takes care of the kitchen and cosmetic line, and his father Renzo Zangrando (whose mother was one from Damos) is in charge of accounting, cultural activities, and relations with the local area. The traditional sense of family ownership, but also hospitality has remained and distinguishes the atmosphere of the center. They are all aware to be somewhat out of this world, but have the dignity and value of their history.

The case of Damos is part of a development model for the mid-altitude mountain that is based on becoming aware of the immense heritage of historical memory. Young people may return from the city, with a different eye after attaining a college degree, with the true challenge being to overcome parochialism and grow in harmony and mutual cooperation.

Acknowledgements The authors convey their sincere gratitude to the stakeholders and experts who accompanied us throughout the entire research process: **Andrea Bartoli**, Founder of Farm Cultural Park, Favara (Sicily); **Antonio Capone**, Prof. of Telecommunications, Politecnico di Milano, Department of Architecture, Department of Electronics, Information and Bioengineering; **Michele Da Pozzo**, Director of the Natural Park of Ampezzo Dolomites; **Stefano Della Torre**, Prof. of Architectural Conservation, Politecnico di Milano, Department of Architecture, Built Environment and Construction Engineering; **Dario Foppoli**, Technical Director of Foppoli Moretta e Associati engineering company, Tirano (Region of Lombardy), and *Via dei Terrazzamenti* project coordinator; **Pietro Guindani**, Fastweb + Vodafone; **Salvatore La Spina**, Mayor of the Municipality of Centuripe (Region of Sicily); **Roberto Laffi**, General Director for Territory and Green Systems, Regione Lombardia; **Pietro Petraroia**, CEO of Cultura Valore S.r.l., Milan; **Serenella Sala**, Founder of Serenella Sala & Associati, Milan; **Marco Scaramellini**, Mayor of the Municipality of Sondrio (Region of Lombardy); **Francesca Canovi**, Deputy Mayor of the City of Sondrio, and Alderman for Production and Events of the City of Sondrio; **Michele Diasio**, Alderman for Tourism, 2026 Olympics, and Sports of the City of Sondrio; **Carlo Mazza**, Alderman for Urban Planning, Environment, and Mobility of the Municipality of Sondrio; **Gianni Verga**, past Alderman for territorial development, Regione Lombardia and City of Milan; **Giacomo Zangrando**, Farmhouse Damos, Valle di Cadore.

In addition to the above-mentioned stakeholders and experts, the Authors would like to thank, for their contribution to the research through various suggestions, trend insights, and best practices: Riccardo Bella, a Milan-based entrepreneur who restored a cabin in a high pasture above the village of Rima (Valsesia) as a family hermitage and hunting hut; Isaac Castellani of the Cà Rossa B&B, Montagna in Valtellina (Valtellina); Giovanna Cicogna, an architect who experienced the most innovative apps for visiting isolated heritage buildings and crossed by bicycle the entire Apennine chain as a form of post-cancer therapy; the Contrada Beltramelli Agritourism, Villa di Tirano (Valtellina) and his architect Daniele Sambrizzi (Valdisotto, Valtellina); the Curt di Clement eco-mobility hotel, Tirano (Valtellina); Clementina Dotti, who, with the National Trust for Italy, opens to the public the historic family mansion located in Pontremoli, a shrinking Apennine small town; Davide Fasolini of Dirupi (Montagna in Valtellina); Maurizio Federici, General Director of the pre-Alpine Province of Varese, who participated in the Valtellina research workshop; Mattia Gurini of Cucina Yugen, Sondrio (Valtellina); the historic winery La Gatta in Bianzone (Valtellina), which welcomed a research workshop session; Armando Lanzetti of the Wine Hotel Retici Balzi, Poggiridenti (Valtellina); Niccolò Lenoci and Federico Romeri of Pascol, Sondrio (Valtellina); the Olmi Agritourism in Ponchiera (Sondrio); Silvia Scieghi and Stefano Canzi of the organic and biodynamic farm Alpes Agia, Albosaggia (Valtellina); Franco Spada, past-mayor of the City of

Tirano; Bruno Tormena, a Prosecco wine producer (Mongarda) whose family vineyards draw the steep slope landscape of Veneto hills; Evaristo Zanolari of ZARCH s.a., Poschiavo, Switzerland, who guided the cross-border workshop experience on the Bernina Unesco Train and across the village of Poschiavo; Sandra Zappella, Director of the Landscape Unit of the Regione Lombardia, who participated in the Valtellina research workshop.

References

Bonfanti, A. (Ed.). (2024). *Borghi avvenire*. Il dossier di Legambiente. https://www.legambiente.it/wp-content/uploads/2021/11/Borghi-Avvenire.pdf. Last accessed 28 Jan 2025

De Varine, H. (2005). *Les racines du future. Le patrimoine au service du développement local*. ASDIC.

Draghi M. (2024) The future of European competitiveness. https://commission.europa.eu/document/download/97e481fd-2dc3-412d-be4c-f152a8232961_en?filename=The%20future%20of%20European%20competitiveness%20_%20A%20competitiveness%20strategy%20for%20Europe.pdf. Last accessed 28 Apr 2025.

European Commission (2024) WHITE PAPER. How to master Europe's digital infrastructure needs?. https://digital-strategy.ec.europa.eu/en/library/white-paper-how-master-europes-digital-infrastructure-needs. Last accessed 28 Jan 2025.

Fondazione per la Sostenibilità Digitale. (2024). What Italians think of the relationship between sustainability and digital. .

IFEL—Istituto per la Finanza e l'Economia Locale. (2024). Piccoli comuni. https://www.fondazioneifel.it/images/PiccoliComuni_marzo24.pdf. Last accessed 28 Jan 2025.

Letta E. (2024). *Much more than a market. Speed, security, solidarity*. https://www.consilium.europa.eu/media/ny3j24sm/much-more-than-a-market-report-by-enrico-letta.pdf. Last accessed 28 Jan 2025.

Small Town Renaissance. A Global Need, Not a Local Dream

Brent D. Ryan and Giovanna Fossa

Beyond Potential, an Emerging Promising Trend

The research approach, supported by a global perspective and data-driven analyses combined with on-site workshops, has been developed through a participatory process. This involved discussions and sharing of challenges, ideas, and visions with local stakeholders as well as international, interdisciplinary experts.

The working hypothesis, focusing on the renaissance of shrinking small towns with historic landscapes, is emerging not only as a feasible opportunity but, in many significant cases, as a contemporary trend. This trend is made possible by the digital transition and driven by new post-COVID lifestyles and forms of tourism that seek quiet, authentic, integrated, identity-rich, and inspiring landscapes. In Italy, these resources are particularly evident due to the micro-identity of its physical and cultural landscapes, positioning the country as a privileged experimental lab for inland territories worldwide.

B. D. Ryan
Department of Urban Studies and Planning, Massachusetts Institute of Technology, Cambridge, MA, USA
e-mail: bdr@mit.edu

G. Fossa (✉)
Department of Architecture, Built Environment and Construction Engineering, Politecnico di Milano, Milan, Italy
e-mail: giovanna.fossa@polimi.it

G. Fossa et al. (eds.), *Small-Town Renaissance*, Research for Development,
https://doi.org/10.1007/978-3-031-89786-3_10

Planning Visions on the Background of Digital Transition

The shared vision for the renaissance of shrinking small towns is not driven by a neo-romantic ideal, but rather as a response to the urbanization process—and the related shrinkage of inner territories—being accepted as irreversible. The global contemporary narrative on human habitats tends to promote a scenario in which cities become increasingly dense, merging with the wild. Urbanization is seen as something to be contained only through land preservation and densification, while inland landscapes are left abandoned, allowing the return of the wild as an environmental victory. However, the abandonment of rural activities such as pastures, cultivated fields, and terraces cannot restore the original ecosystem wilderness in historically humanized landscapes (a Rousseau-esque myth). In fact, abandonment leads to environmental degradation, loss of biodiversity, and increased hydrogeological risks, as well as the irreversible loss of built and cultural heritage.

The alternative vision emerging from the research promotes regional polycentrism, rooted in a re-inhabited countryside with small towns marking the rural landscape. This model supports innovative, contemporary, and even high-tech agriculture, yet still deeply connected to local places and communities, as opposed to the super-industrialized agriculture demanded by megacities. In this context, the 2020 exhibition "Countryside: The Future" by Koolhaas (AMO, Koolhaas, 2020), set during the pandemic, proved prophetic. The vision integrates tourism with culture and production—rural, artisanal, and manufacturing—creating a productive, living, and integrated landscape. Climate change urgently necessitates this shift in order to prevent hydrogeological risks and escape urban heat islands, all while exploring the environmental quality of life offered by small towns and villages situated at higher altitudes than metropolitan plains and coastlines.

Of course, both models coexist in the complex dynamics of ongoing territorial development, but a rebalancing should be pursued, especially to build resilience against global crises.

The pandemic highlighted the shortcomings of overly rigid, transit-oriented development, especially considering the health risks associated with congested transit systems. It has prompted a rediscovery of the resilience found in small-town networks and highlighted the value of vacation homes and family-owned rural properties as temporary—often year-round—living solutions, supported by new telecommunication technologies.

Technology and connectivity have long shaped settlement patterns and the way we experience and interact with territories. Wireless bridges and satellite coverage have made digital accessibility pervasive, even in mountainous and remote inland areas. Remote working opportunities are now available even in isolated villages, and new possibilities emerge through e-commerce of local products, innovative high-tech agriculture, and drought management, all within increasingly meaningful landscapes. However, telecommunication technologies must be supported by public policies that ensure digital accessibility for all, not just for market-driven areas, to make internet services sustainable for people in sparsely populated regions.

Post-Covid Lifestyles & Experience Tourism: A New Culture Sensibility

The widespread upgrade of digital accessibility—achieved in Italy with the extension of broadband infrastructure during the pandemic and globally through today's satellites—provided the *conditio sine qua non* for the renaissance of small towns. However, it was the lockdown experience itself that marked a turning point, reversing the trend of decline in many villages, as evidenced by local data on both population and tourism.

The pandemic made us rediscover the outdoors, prioritizing health and wellbeing, with new habits of living and working integrated with leisure and innovative, sustainable forms of tourism. Remote and smart working became institutionalized, free from previous prejudices.

For the past 20 years, the rise of experiential tourism has highlighted the crisis in traditional mass tourism (Pine, Gilmore, 1999). The pandemic reinforced this shift: off-season experiential tourism and smart/remote working now express converging demands, seeking the same types of identity-rich, uncrowded places where people can interact with the local community and reconnect with the rhythms of nature.

This new vision of the "South" is independent of geographic latitude. Local pioneers include organic farmers, agritech researchers, chefs and winemakers innovating site-specific food and wine, young shepherds welcoming tourists instead of keeping them away as in the past, eco-friendly hoteliers, shelter managers opening year-round, new mountain guides focused not on climbing but on local history, culture, flora, and fauna. International trendsetters include digital nomads, mindfulness gurus, and strategic coaches. It is a shared global cultural sensibility emerging worldwide—echoes of the hippie movement and yoga influences, fueled also by cultural exchanges among extreme climbers from around the world.

Inner identity landscapes are being rediscovered as creative spaces where ideas flourish, and people find themselves. These are regenerative landscapes, even before they are landscapes to be regenerated. The notion of big cities as privileged creative ecosystems (Landry, 2012) has its limits due to congestion and overtourism. For a place to inspire, it must be peaceful—just as historical small towns and villages are.

Moreover, the concept of an "inspiring place" is evolving from a static, post-romantic vision toward a dynamic, post-pandemic one. Inspiration for creativity and self-regeneration is now found through walking, rather than sitting in a "room with a view".[1] The creative perspective is shifting, transforming the traditional grand tour painting views in a contemporary way while maintaining the "en plein air" atmosphere of the Impressionists.

The pandemic taught us the vital importance of outdoor walking, particularly in the context of an aging population, for preventing major health issues such as cardiovascular problems and bone fragility.

[1] "A Room with a View", novel by E. M. Foster, 1908, which inspired the same-name movie directed by J. Irons in 1985.

The new scale of walkability is becoming territorial, connecting historical cores with regional hiking paths, supported by sustainable multimodal mobility options (train, bike, boat, cableway, etc.). An additional advantage is the ability to walk comfortably even in winter, thanks to ongoing climate change.

Walking routes are emerging as strategic territorial spines for regeneration. Historical pilgrim paths, a medieval network crossing the rural landscapes of Europe, are now being rediscovered for experiential tourism, sustainable sports activities, mindfulness, or counseling programs. These paths support synergetic services such as agritourism, food and wine production, and continuous innovation rooted in the rich tradition of historical landscapes. This leads to excellence in both products and the architectural experience offered. At times, the quality achieved can compete in the sophisticated international luxury market, but what is crucial is always guaranteeing the quality of the experience for a wide range of target audiences. The recent global passion for vineyards, which is sweeping contemporary culture, plays a positive role in driving many inner landscapes, creating new precious value chains.

Walking routes in inland territories can also be used for educational and research programs, as demonstrated during the traveling workshops that featured the research subject of this book. This mode follows the contemporary evolution of the link between research and inspiring places, which drove the location choices for research centers, corporate retreats, and learning centers in the past century.[2] A new model of walking research is required by contemporary sensibility and the post-COVID reinterpretation of the classical holistic synthesis of *mens sana in corpore sano*.

An Alternative Mountain

Orographic complexity characterizes many shrinking territories, creating challenges for access and mobility within and between settlement cores and rural areas. Beyond widespread internet connectivity, today technological innovations are physically expanding the horizons of isolated valleys, enhancing opportunities for both residents and tourists. Advanced drones now address last-mile logistics by reaching upper slopes from valley-floor hubs. E-bikes allow users to cover previously impassable distances and elevations, offering increasingly popular routes along white roads and rural or forest paths. Similarly, new e-skis for ski mountaineering increase climbing speed while reducing physical exertion, significantly expanding alpine skiing options. These innovations are particularly appealing for older populations, extending sports participation beyond just the young.

A new, sustainable mountain vision is emerging—one free from ski lifts and invasive air infrastructures, such as power lines (now increasingly buried), and car traffic (which is being restricted). Accessibility is instead limited to cars, trains, or

[2] See for example the Rockefeller Foundation in Bellagio, Lake of Como, opened in 1959.

cable cars only up to entry points for infrastructure-free zones dedicated to walkability and sustainable sports. In this context, isolated valleys and villages that have preserved authentic and integrated landscapes have a strategic advantage. They can avoid abandonment by offering tranquil atmospheres, unique combinations of sports and meditation, and meaningful interactions with local culture and nature—experiences that mainstream tourist destinations cannot provide. Sustainable mobility becomes a key part of the tourist experience and, at times, even the destination itself. Slow mobility options include active recreation, sports, or scenic passive transport such as Alpine trains and sleigh rides. Various experiments with these concepts are already underway in smaller Alpine villages.[3]

Heritage and Identity

The fundamental resource for the renaissance of small towns and villages is the immense richness of their historical landscapes, which remain underexplored. This includes built heritage but, most importantly, intangible cultural heritage—such as local communities themselves, which are at risk.

One of the main obstacles to enjoying built heritage in inner areas has traditionally been its dispersion across vast territories. However, widespread digital accessibility now offers new opportunities to visit even remote little churches and minor monuments through dedicated apps that provide access on demand without requiring on-site staff.

Additionally, digital technologies enable preventive and scheduled maintenance by introducing sensors to monitor structural and climatic conditions. For historical building restoration and adaptive reuse, new technologies and techniques are available that are compatible not only with preservation but also with achieving advanced energy performance. The key challenge remains how to create a coherent system from dispersed heritage while engaging the local community.

The cultural district model[4] proved fruitful over the past 20 years. Today, thanks to innovative apps and remote control technologies, it is possible to go further by integrating built and cultural heritage with sports and sustainable mobility routes within the same value chain. Customizable and real-time integrated itineraries for experiential tourism can be created, offering space-time geographies resilient to weather changes and personal preferences. This intersectoral system serves both residents and tourists, creating new ways to experience the territory for living, working, and tourism.

---

[3] See for example the recent experiment of Montespluga (Sondrio) and the precursor experience of Belalp (Switzerland).

[4] The "cultural district" is a local development model emerged in Italy around the year 2000, which leverages on heritage-based innovative projects, integrating different institutional heritage sectors and engaging the territory. See Della Torre, 2015.

Cultural district experiences have demonstrated that engaging the local community requires a focus on selected tangible projects where various public and private institutions can align their interests within the framework of collective benefit. These projects should be strategic in nature, triggering spontaneous regeneration in line with contemporary planning approaches that combine a shared vision with flagship projects. This desired public-private partnership can be realized in a participatory way with local people, the so-called P4 model: Public-Private-People Partnership (Boniotti, 2023).

Walking routes are ideal strategic structures for regenerating both spaces and communities. The encounter between heritage and contemporary art has also proven to be a powerful tool for revitalization and rebranding, often engaging local crafts and attracting international audiences.

Heritage is closely intertwined with the physical and cultural identity of places—the "spirit of the place". This identity encompasses landscapes, built and cultural heritage, languages, toponyms, biodiversity, and endemic species. The concept of "glocal" (Bassetti, 2015) summarizes the essential balance between participation in international networks, supported by upgraded internet connections, and preserving local specificities, identity resources, and historical and cultural roots.

This is not a simple issue, as it requires both individual and collective awareness of being custodians and bearers of historical and cultural treasures, as well as fostering a cohesive community mindset (Manfredi, Costi, 2023). It is essential for national and supranational public policies to adopt site-specific and culture-specific approaches rather than rigid, universal performance parameters.

Micro-Entrepreneurship

The concept of a productive district[5] complements that of a cultural district, allowing landscapes historically shaped by rural and manufacturing production to continue or return to productivity in innovative, yet site-specific ways. Innovation, particularly in the digital transition, must be rooted in the local community and its landscape.

It is strategic to train young people to develop a pride of belonging, without falling into either conceit or victimhood. This involves fostering a mindset that counters parochialism and encourages innovative micro-entrepreneurship. While exogenous private investment can be valuable, it is most effective when it triggers a process that revitalizes local micro-entrepreneurship. In contrast, the purchase of entire hamlets by multinationals may preserve buildings from ruin but risks

[5] The "productive distric" model emerged in Italy in the '80s promoting small medium enterprise chains which feature identity regional production. More recently (2009) Giancarlo Corò and Stefano Micelli observed their innovation dynamics and relaunched their cultural value and global competitiveness in the research "I nuovi distretti produttivi", Marsilio, 2009; for inner territories, the focus is on artisan know-how and small business, see Micelli, 2011.

alienating the community, transforming an authentic territory into a museum or playground.

Villages are saved by a combination of community and micro-entrepreneurship, which in Italy often aligns with family communities and family-run businesses. This combination forms district networks, where supply chains—sometimes starting with crafts and extending to contemporary art—are innovating, developing e-commerce, and integrating events, transportation, and mobility solutions.

Micro-entrepreneurship in a globalized market requires investment and incentives from an innovative public administration.[6] This approach could also address the critical issue of services in extended territories with scattered settlements. Micro-entrepreneurship, especially in service production, holds great potential—particularly in the development of innovative people-centered services that require youthful energy and fresh ideas. These could be run as private enterprises in partnership with the public sector.

Rather than traditional, localized public services, it would be more appropriate to promote on-demand, home-based services supported by technology and widespread digital accessibility. These services could be integrated with hubs for sustainable mobility and innovative last-mile logistics.

The New "Place to Be"

Hyperconnectedness can enhance hyperlocal experiences and foster the formation of new communities to complement or revitalize traditional ones. These might include smart workers, digital nomads, and alternative sports enthusiasts.

To encourage repopulation, in addition to retaining and attracting permanent residents—the essential builders of landscapes that require long-term care—it is worth explicitly targeting temporary residents based on different life and work phases. These could include remote workers, families with children, young innovative entrepreneurs, retirees, and long-term tourists. There are reports of young couples returning to their native villages after studying or working elsewhere, as well as metropolitan residents moving to villages as a lifestyle choice, finding a sense of belonging by election.

With digital accessibility, choosing a small town as a place to live has become a cultural decision rather than one driven by employment. Multinational companies now hire remotely, making job location less relevant. Instead, the decision hinges on culture, psychology, and an appreciation for the local community or family of origin. Social rootedness is emerging as a valuable concept to rediscover in today's fluid society (Bauman, 2020).

[6] For the public sector innovation challenge see "The Creative Bureaucracy Festival", launched by Charles Landry in Berlin in 2025.

The primary resource of small towns is the culture of their local communities, deeply rooted in the history and natural environment of the place. This landscape must remain "vital"—not merely consumed or passively enjoyed but actively shaped and built collectively. While there is genuine hope for revitalization, it requires bold and courageous policies to succeed.

Acknowledgements Authors thank the MISTI Program for having planted the seeds of this shared research path, through the MIT Italy/Politecnico di Milano Rocca Project.

References

AMO, Koolhaas R. (2020). *Countryside. A report*. Taschen.
Bassetti, P. (2015). *Svegliamoci italici! Manifesto per un futuro glocal*. Marsilio.
Bauman, Z. (2020). *Liquid modernity*. Polity.
Boniotti, C. (2023). The public–private–people partnership (P4) for cultural heritage management purposes. *Journal of Cultural Heritage Management and Sustainable Development, 13*(1), 1–14. https://doi.org/10.1108/JCHMSD-12-2020-0186
Della Torre, S. (2015). Shaping tools for built heritage conservation: From architectural design to program and management. Learning from "Distretti culturali". In K. Van Balen & A. Vandesande (Eds.), *Community involvement in heritage* (pp. 93–101). Garant.
Landry, C. (2012). *The creative city*. Earthscan.
Manfredi, F., & Costi, D. (2023). *Community regeneration masterplan*. Springer.
Micelli, S. (2011). *Futuro artigiano*. Marsilio.
Pine, B. J., & Gilmore, J. H. (1999). *The experience economy*. Harvard Business Review Press.

The manufacturer's authorised representative in the EU is Springer Nature Customer Service Centre GmbH, Europaplatz 3, 69115 Heidelberg, Germany. If you have any concerns regarding our products, please contact ProductSafety@springernature.com

Printed and bound by CPI Group (UK) Ltd, Croydon, CR0 4YY
01/10/2026
02225961-0008